Developing Numeracy
MEASURES, SHAPE & SPACE
ACTIVITIES FOR THE DAILY MATHS LESSON

year

Hilary Koll & Steve Mills

A & C BLACK

Contents

Introduction 4

Measures

Paperclip measuring	estimate, measure and compare length using non-standard units	6
10 centimetre Clive	estimate, measure and compare length using standard units	7
Centimetres or metres?	suggest suitable units for measuring length	8
Getting equipped	suggest suitable equipment for measuring length	9
Metre stick measuring	read simple scales for length	10
Fishing lines	use a ruler to draw and measure lines to the nearest cm	11
Snail trails	use a ruler to draw and measure lines to the nearest cm	12
Balancing act	estimate, measure and compare mass using non-standard units	13
A fine balance	estimate and compare the mass of objects	14
Mass mix-up	estimate the mass of objects as less than a kilogram/ a kilogram or more	15
Grams or kilograms?	suggest suitable units for measuring length	16
Guessing game	estimate the mass of objects as less than/more than/about a kilogram	17
Weighty problems	read simple scales for mass	18
Which holds more?	estimate and compare the capacities of containers	19
Bring and buy	estimate the capacity of objects as less than a litre/a litre or more	20
Millilitres or litres?	suggest suitable units for measuring capacity	21
Find it	estimate and compare the capacities of containers using litres	22
Waterworld	read simple scales for capacity	23
Gemma's calendar	know and order the months of the year	24
Know your months	know and order the months of the year	25
How long?	suggest suitable units to estimate or measure time	26
What's the time, Mr Wolf?	read the time to the hour, half hour or quarter hour on an analogue clock	27
Time match	read the time to the hour, half hour or quarter hour on an analogue clock	28
Domino time	read the time to the hour, half hour or quarter hour on digital and analogue clocks	29
Quiz time	read the time to the hour, half hour or quarter hour on a digital clock and understand the notation	30

Shape and space

Shape collector	use the mathematical names for common 3-D shapes	31
Shape puzzle	use the mathematical names for common 3-D shapes and recognise their features	32
Feely bag fun	use the mathematical names for common 3-D shapes and recognise their features	33
Shape hunt	recognise and describe the features of 3-D shapes	34
Skeletons	make 3-D shapes and describe their features	35
True or false?	relate solid shapes to pictures of them	36
Shape stories	use the mathematical names for common 2-D shapes	37
Wrapping paper	use the mathematical names for common 2-D shapes	38
Shape cards	sort common 2-D shapes and describe their features	39
Picture perfect	make and describe shapes, pictures and patterns	40
Pentagon patterns	make and describe shapes, pictures and patterns	41

Tile teasers: 1	make and describe shapes, pictures and patterns	42
Tile teasers: 2	make and describe shapes, pictures and patterns	43
Animal symmetry	recognise symmetrical and non-symmetrical shapes	44
Paper shapes	begin to recognise line symmetry	45
Spaceman Sam	sort symmetrical and non-symmetrical shapes	46
Road signs	draw a line of symmetry	47
Dotty's friends	draw and make symmetrical patterns	48
Kite flying	draw and make symmetrical patterns	49
Fruit shop	use mathematical vocabulary to describe position	50
Toy shop	use mathematical vocabulary to describe position	51
Which letter?	use mathematical vocabulary to follow directions	52
Safari survival!	use mathematical vocabulary to follow directions	53
Cat and mouse	use mathematical vocabulary to follow directions	54
Star maze	give instructions for moving along a route and round corners	55
Robot rotations	recognise clockwise and anticlockwise, half and quarter turns	56
On the spot	follow half and quarter turns	57
Right angle gobbler	recognise and find right angles	58
Barmy borders	describe movements and understand angle as a measurement of turn	59
Tick the turn	describe movements and understand angle as a measurement of turn	60
Name the turn	describe half and quarter turns (right angles)	61
Beautiful borders	describe what is happening in a repeating pattern	62
Acrobats	describe whole, half and quarter turns	63
Answers		64

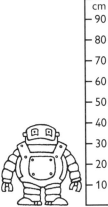

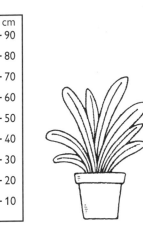

Reprinted 2002 (twice), 2003, 2006
Published 2001 by A & C Black Publishers Limited
38 Soho Square, London W1D 3HB
www.acblack.com

ISBN-10: 0-7136-5877-0
ISBN-13: 978-0-7136-5877-4

The authors and publishers would like to thank Madeleine Madden and Corinne McCrum for their advice in producing this series of books.

A CIP catalogue record for this book is available from the British Library.

A & C Black uses paper produced with elemental chlorine-free pulp, harvested from managed sustainable forests.

Printed in Great Britain by Caligraving Ltd, Thetford, Norfolk.

Introduction

Developing Numeracy: Measures, Shape and Space is a series of seven photocopiable activity books designed to be used during the daily maths lesson. They focus on the fourth strand of the National Numeracy Strategy *Framework for teaching mathematics*. The activities are intended to be used in the time allocated to pupil activities; they aim to reinforce the knowledge, understanding and skills taught during the main part of the lesson and to provide practice and consolidation of the objectives contained in the framework document.

Year 2 supports the teaching of mathematics by providing a series of activities which develop essential skills in measuring and exploring pattern, shape and space. On the whole the activities are designed for children to work on independently, although this is not always possible and occasionally some children may need support.

Year 2 encourages children to:

- use and begin to read the language of measure, shape, space and time;
- estimate, measure and compare lengths, masses, capacities and time, and to suggest suitable units and equipment for such measurements;
- read a simple scale to the nearest labelled division, and to draw and measure lines to the nearest centimetre;
- recognise 3-D and 2-D shapes and describe some of their properties;
- begin to recognise line symmetry;
- use mathematical language to describe position, direction and movement.

Extension

Many of the activity sheets end with a challenge (**Now try this!**) which reinforces and extends the children's learning, and provides the teacher with the opportunity for assessment. On occasion, you may wish to read out the instructions and explain the activity before the children begin working on it. The children may need to record their answers on a separate piece of paper.

Organisation

Very little equipment is needed, but it will be useful to have available rulers, scissors, coloured pencils, interlocking cubes, counters, solid shapes, dice and small mirrors. You will need to provide paperclips or matchsticks for page 6, string for page 7, a set of pens and pencils for page 11, balance scales and a range of objects for page 13, dotty triangular paper for page 43, coin for page 54, and a paper or card circle for page 58.

Children should have access to measuring equipment to give them practical experience of length, mass and capacity.

To help teachers to select appropriate learning experiences for the children, the activities are grouped into sections within each book. However, the activities are not expected to be used in that order unless otherwise stated. The sheets are intended to support, rather than direct, the teacher's planning.

Some activities can be made easier or more challenging by masking and substituting some of the numbers. You may wish to re-use some pages by copying them onto card and laminating them, or by enlarging them onto A3 paper.

Teachers' notes

Very brief notes are provided at the foot of each page giving ideas and suggestions for maximising the effectiveness of the activity sheets. These can be masked before copying.

Structure of the daily maths lesson

The recommended structure of the daily maths lesson for Key Stage 1 is as follows:

Start to lesson, oral work, mental calculation	5–10 minutes
Main teaching and pupil activities (*the activities in the* **Developing Numeracy** *books are designed to be carried out in the time allocated to pupil activities*)	about 30 minutes
Plenary (*whole-class review and consolidation*)	about 10 minutes

Whole-class warm-up activities

The following activities provide some practical ideas which can be used to introduce or reinforce the main teaching part of the lesson.

Measures

Make a metre
Call out measurements and ask the children to tell you how many more centimetres are needed to make a metre, for example: *20 centimetres; how many more?*

Measurement song
To the tune of *London's Burning* sing an 'echo' song where you sing one line and the children repeat it. These verses can be used to help children remember the relationships between units of length, mass and capacity.

A hundred centimetres	(*hundred centimetres*)
makes a metre	(*makes a metre*)
This is length	(*this is length*)
Now we know it	(*now we know it*)
A thousand grams	(*thousand grams*)
is a kilogram	(*is a kilogram*)
This is mass	(*this is mass*)
Now we know it	(*now we know it*)
A thousand millilitres	(*thousand millilitres*)
is a litre	(*is a litre*)
Capacity!	(*capacity*)
Now we know it	(*now we know it*)

Estimating activities
Ask the children to estimate the length, mass or capacity of objects around the classroom or use household groceries (with actual measures masked over). Encourage them to suggest a range within which the measurement might fall, for example between 1 and 2 metres. Once the class have agreed a range, the children can test their estimates by measuring.

Mass guess
Provide two boxes of similar size but different weights and ask the children questions such as: *Do you think these boxes weigh the same? Can you tell, just from looking, which box is heavier?*

Show a third box which is larger and lighter than the other two. *Which do you think is heaviest? Are big things always heavier?*

Invite several children to feel the boxes and continue: *Can anyone think of something that is very large but very light? Or something that is small but very heavy?* Discuss the children's suggestions.

Time game
Sit the children in a horseshoe and agree which will be the top end, and which the bottom end. Select children randomly (one at a time) and show them a time on an analogue clock. If they say the time correctly to the nearest 15 minutes, they move to the top. Adjust the level of questions to suit individuals.

Practise counting round the horseshoe in fives. Turn the minute hand of the clock as you do so and emphasise the fifteen, thirty and forty-five.

Shape and space

2-D shape
On the board, draw three large boxes labelled pentagon, hexagon and octagon. Discuss the definitions of these shape words and select children to come to the board and draw the relevant shape in each box. This should encourage children to draw irregular shapes (those that do not have equal sides and equal angles), and so gain a fuller appreciation of these shape names. Paper shapes could be shown and the children could stick them in the correct position on each box.

Name the shape
Hold up a shape and ask the children to name it and/or find its written name from some shape name cards.

Pick a shape but do not show it to the children. Describe it and ask them to guess the shape name. They can ask three questions about it before guessing the shape, for example *Does it have three sides?*

Hidden shapes
Find a given shape in objects which are in the classroom or can be seen from the classroom window. Ask: *Can anyone find a cuboid in our classroom?* This can also be played as an 'I Spy' game.

Clockwise and anticlockwise
Ask the children to stand up and close their eyes. Call out instructions, such as: *Make a quarter turn clockwise* or *Make a full turn anticlockwise*. Encourage the children to turn slowly and not to look at the others.

Turtle moves
Using programmable toys, such as Roamer or Pip, ask the children to suggest how to programme the toy to move it from one point to another. Encourage them to use the language of movement and direction.

Paperclip measuring

- **Fill in the chart below.**

  [Estimate] **how many paperclips you think you will**

  need. Now [measure] **each length with paperclips.**

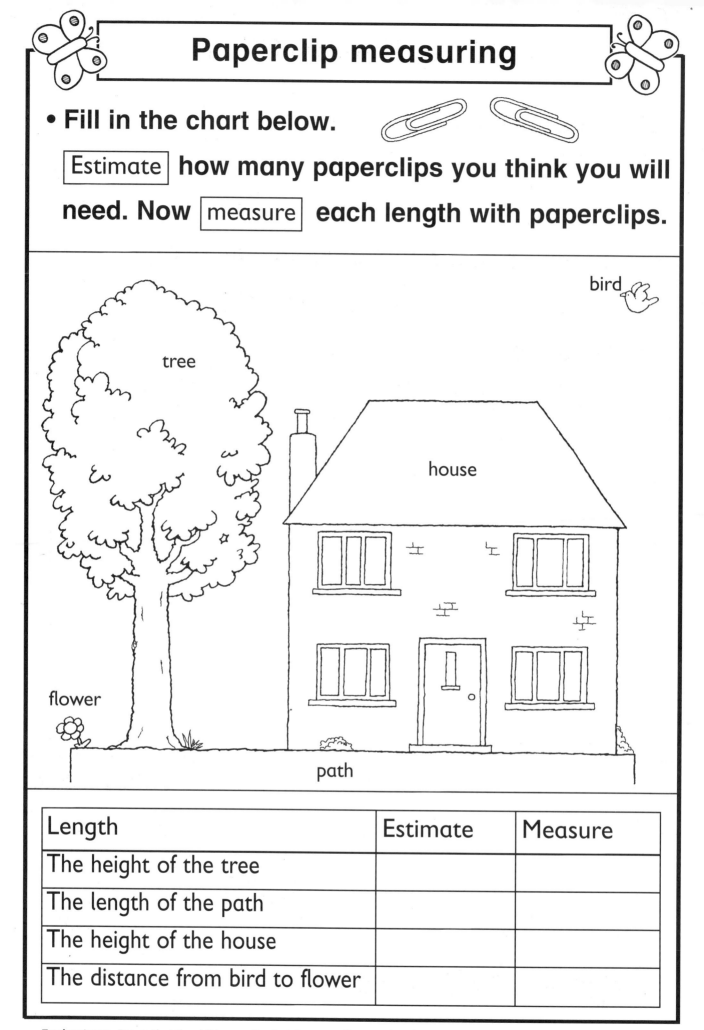

bird

tree

house

flower

path

Length	Estimate	Measure
The height of the tree		
The length of the path		
The height of the house		
The distance from bird to flower		

Teachers' note Ensure that the children realise that the paperclips must be placed in a line end-to-end to gain an accurate measurement. Matchsticks or a similar resource could be used in place of paperclips. For a further extension, the children could use centimetre cubes to begin to measure and record the lengths of lines using centimetres.

Developing Numeracy
Measures, Shape and Space
Year 2
© A & C Black

10 centimetre Clive

Clive the caterpillar is [cm = centimetres]

[10 centimetres] long.

- **Cut some string exactly the length of Clive.**

- **Use the string to measure four objects.**

- **Draw the objects on the chart.**

Shorter than 10 cm	Longer than 10 cm

- **Use your string to measure these caterpillars.**

- **Colour the one that is exactly 10 cm long.**

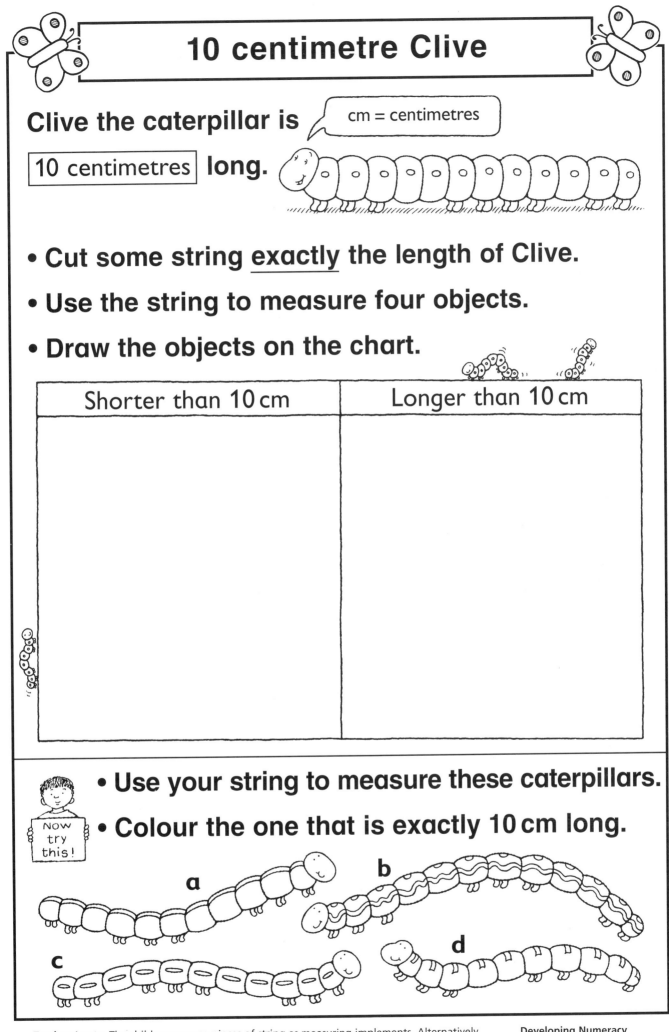

a

b

c

d

Teachers' note The children can use pieces of string as measuring implements. Alternatively, some children could be given rulers to use. References to 10 cm can be masked and the activity carried out for any length, for example 1 metre. The children can be asked to estimate first and fill in the chart before measuring.

Developing Numeracy
Measures, Shape and Space
Year 2
© A & C Black

Centimetres or metres?

- **You can use** | centimetres | **or** | metres | **to measure these things. Write which is best.**

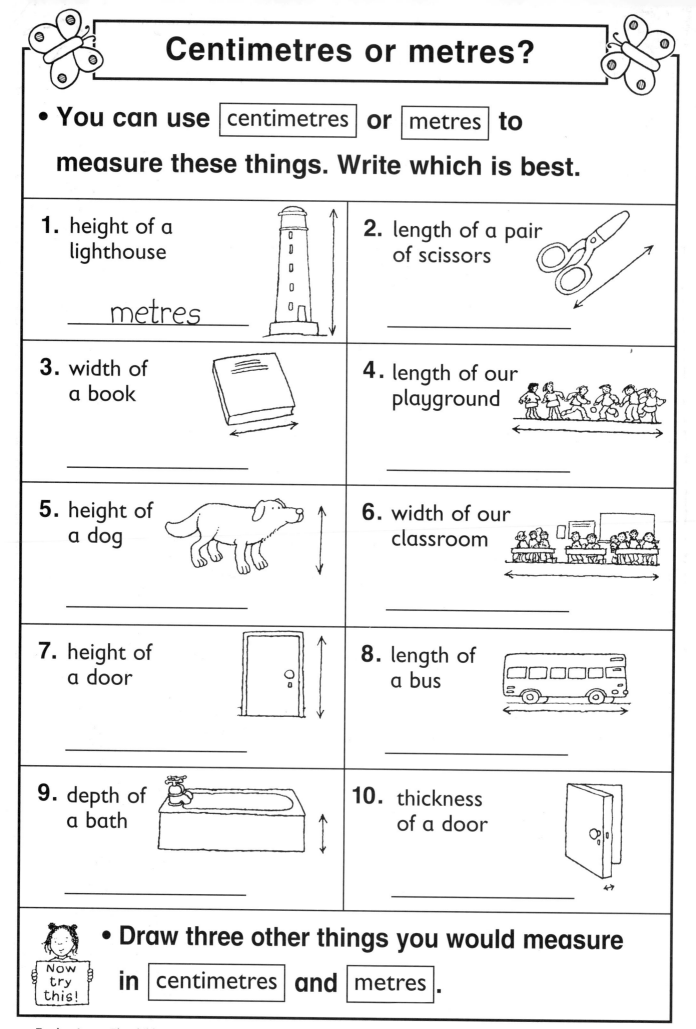

1. height of a lighthouse

   ___metres___

2. length of a pair of scissors

   _____

3. width of a book

   _____

4. length of our playground

   _____

5. height of a dog

   _____

6. width of our classroom

   _____

7. height of a door

   _____

8. length of a bus

   _____

9. depth of a bath

   _____

10. thickness of a door

   _____

- **Draw three other things you would measure in** | centimetres | **and** | metres |.

Teachers' note The children can compare answers with a partner. Ensure that the children appreciate that the terms length, distance, height, width, depth and thickness are all lengths and can be measured using metres and centimetres.

Developing Numeracy
Measures, Shape and Space
Year 2
© A & C Black

Getting equipped

- **Look at the four pieces of equipment.**
- **Write which one you would use to measure these.**

Follow this trail.

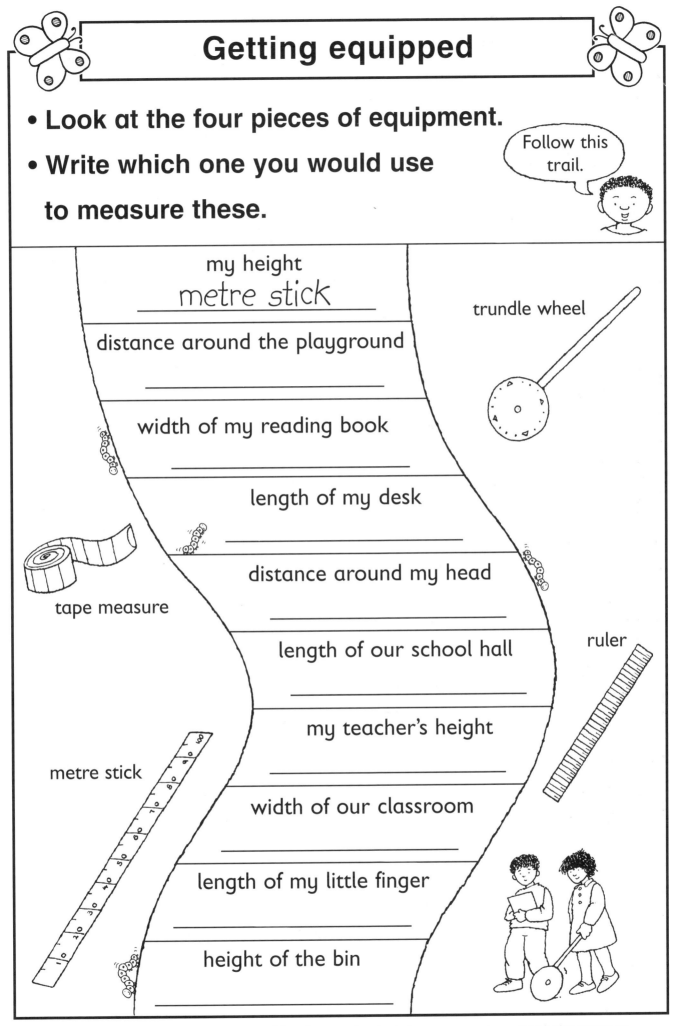

my height

metre stick

distance around the playground

width of my reading book

length of my desk

distance around my head

length of our school hall

my teacher's height

width of our classroom

length of my little finger

height of the bin

trundle wheel

tape measure

metre stick

ruler

Teachers' note Introduce the children to metre sticks, rulers, trundle wheels, tape measures, etc. at the start of this lesson. Ensure that the children understand how they are used and can appreciate that some pieces of equipment are used for measuring longer lengths whilst others are used for shorter lengths, and some for lengths that are not straight.

**Developing Numeracy
Measures, Shape and Space
Year 2
© A & C Black**

Metre stick measuring

- **Write the** heights **in** centimetres .

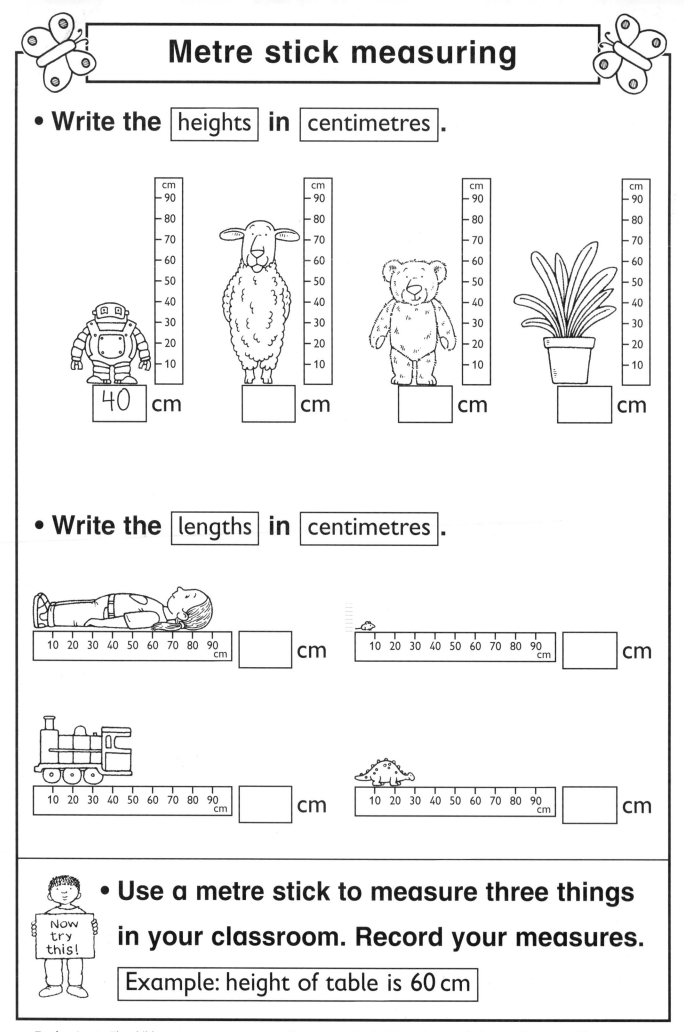

40 cm

cm

cm

cm

- **Write the** lengths **in** centimetres .

cm

cm

cm

cm

- **Use a metre stick to measure three things in your classroom. Record your measures.**

Now try this!

Example: height of table is 60 cm

Teachers' note The children can compare answers with a partner. Remind them that 'cm' is short for centimetres. The pictures could be masked to provide a flexible resource and, as an extension, new items drawn that are between multiples of ten centimetres, for example 85 cm.

Developing Numeracy
Measures, Shape and Space
Year 2
© A & C Black

10

Fishing lines

- **Use a ruler to measure the** $\boxed{\text{lengths}}$ **of the gnomes' fishing rods in** $\boxed{\text{centimetres}}$ **.**

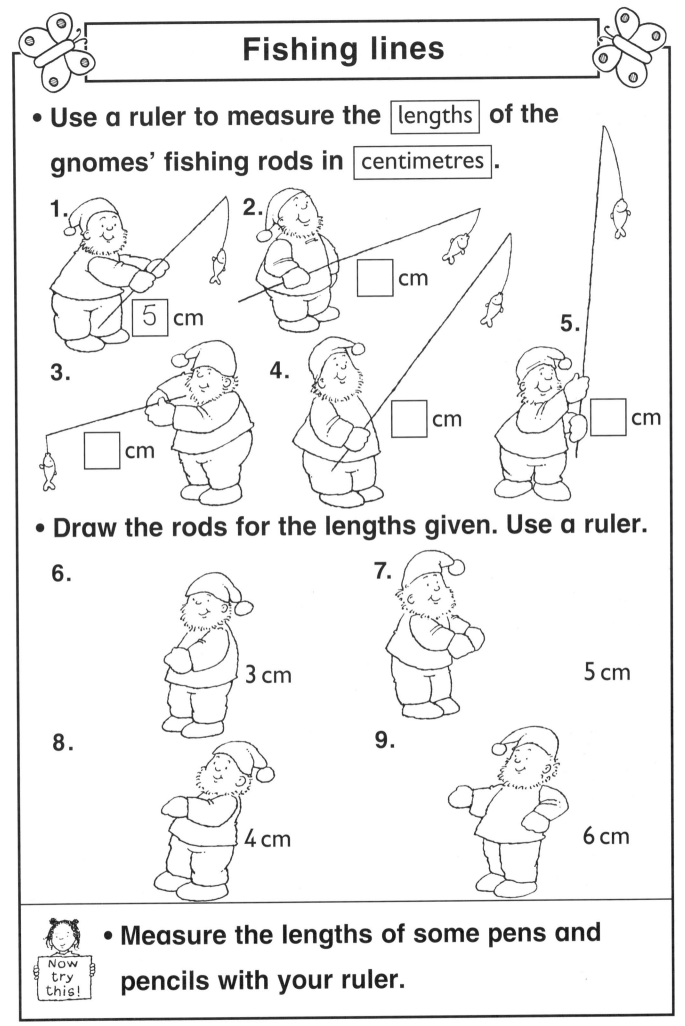

1. $\boxed{5}$ cm

2. ☐ cm

3. ☐ cm

4. ☐ cm

5. ☐ cm

- **Draw the rods for the lengths given. Use a ruler.**

6. 3 cm

7. 5 cm

8. 4 cm

9. 6 cm

- **Measure the lengths of some pens and pencils with your ruler.**

Now try this!

Teachers' note Children often misuse rulers that have a gap between the end of the ruler and where the scale starts. Show the children where the end of the line must be positioned for the type of ruler being used. Explain that the scale on a dead-end rule begins at the very end of the ruler.

Developing Numeracy
Measures, Shape and Space
Year 2
© A & C Black

11

Snail trails

- **Use a ruler to measure the** lengths **of the snail trails in** centimetres **.**

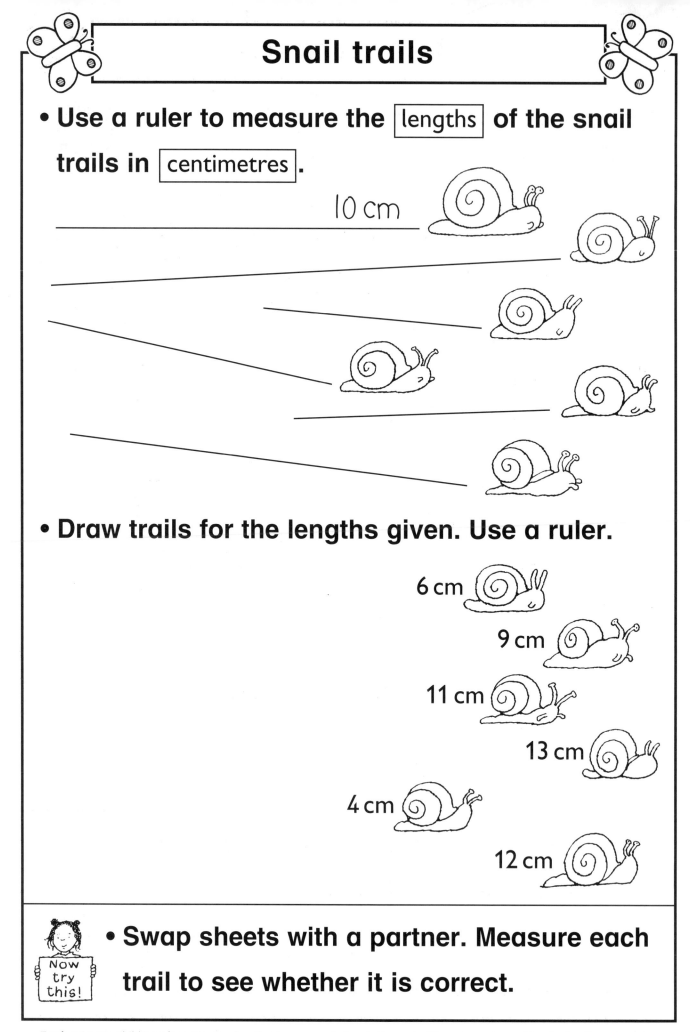

10 cm

- **Draw trails for the lengths given. Use a ruler.**

6 cm

9 cm

11 cm

13 cm

4 cm

12 cm

- **Swap sheets with a partner. Measure each trail to see whether it is correct.**

Now try this!

Teachers' note Children often misuse rulers that have a gap between the end of the ruler and where the scale starts. Show the children where the end of the line must be positioned for the type of ruler being used. Remind the children that the scale on a dead-end rule begins at the very end of the ruler.

Developing Numeracy
Measures, Shape and Space
Year 2
© A & C Black

Balancing act

- **Write how many cubes** balance **each object.**

You need:
- a balance
- interlocking cubes
- five objects

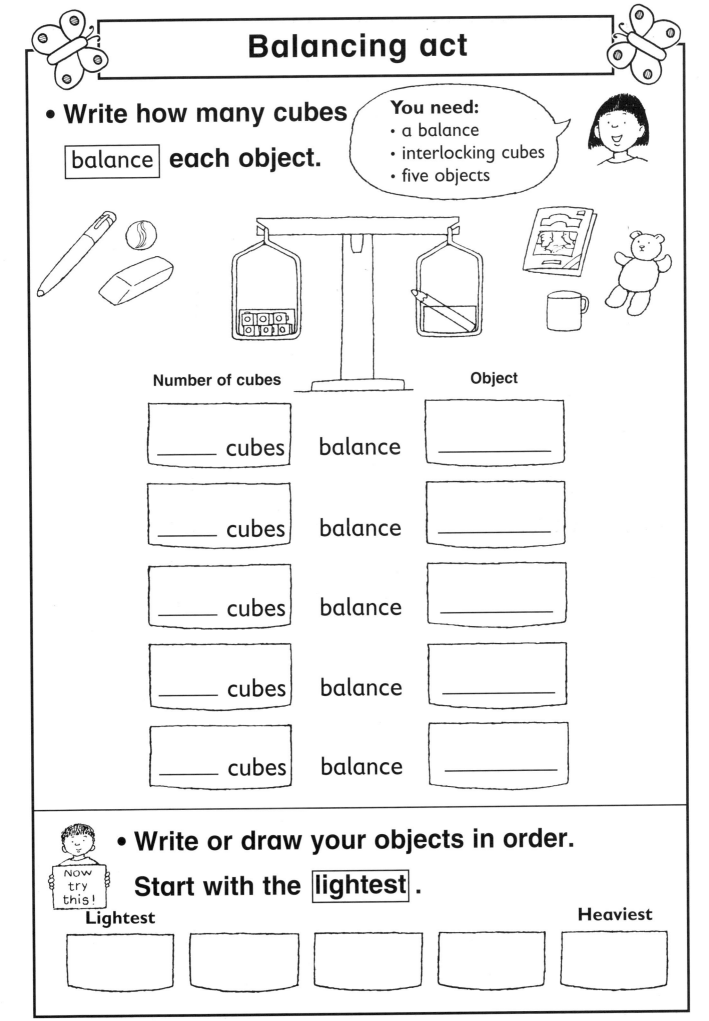

Number of cubes		Object
_____ cubes	balance	_____
_____ cubes	balance	_____
_____ cubes	balance	_____
_____ cubes	balance	_____
_____ cubes	balance	_____

- **Write or draw your objects in order.**

  Start with the lightest **.**

Now try this!

Lightest   **Heaviest**

Teachers' note The children will need to work in pairs or small groups. Provide a range of light objects to balance, for example pen, beaker, ruler, rubber, book, spoon, teddy bear, toy. Encourage the children to use this information to guess how many cubes balance other objects, and to make comparisons between objects by asking which is heavier/heaviest, lighter/lightest, etc.

**Developing Numeracy
Measures, Shape and Space
Year 2
© A & C Black**

13

A fine balance

- ## Tick the heavier thing in each pair.

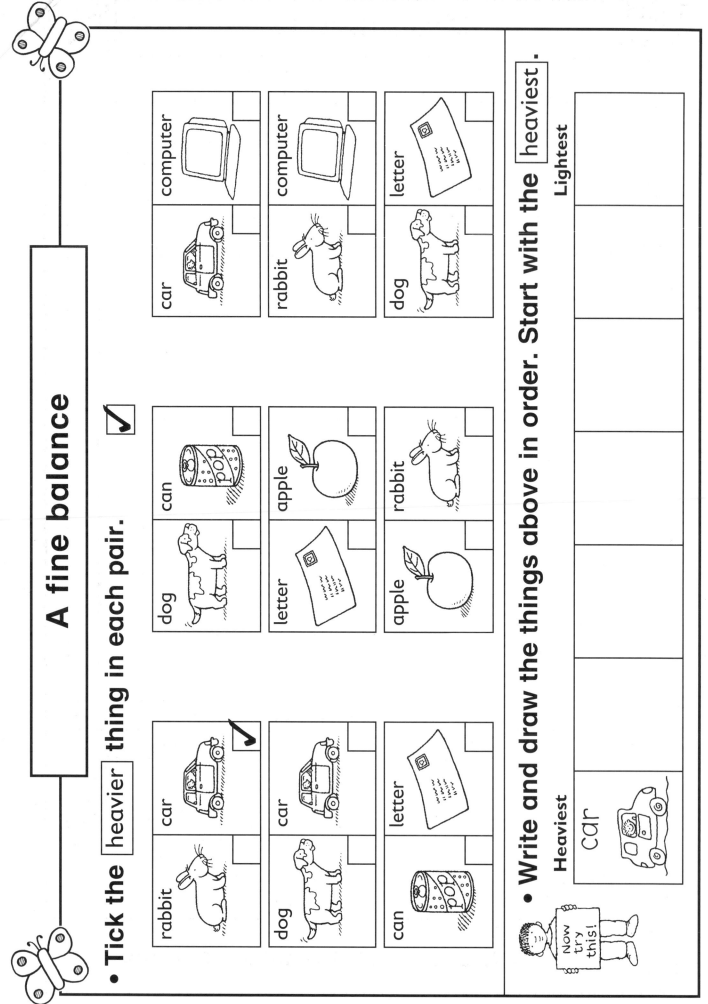

- ## Write and draw the things above in order. Start with the heaviest .

Heaviest

car

Now try this!

Lightest

Teachers' note The children can work in pairs to discuss the extension activity. This sheet could be used following a practical lesson using balance scales.

Developing Numeracy
Measures, Shape and Space
Year 2
© A & C Black

14

Mass mix-up

- ## Work with a partner.

- ## Draw the things on
  ## the correct shelf.

Remember a bag of sugar balances a kilogram weight.

↑ Less than a kilogram ↑

sugar

↑ A kilogram or more ↑

kitchen rolls
rice
smarto
potatoes
jelly
sugar
lemonade
cornflakes
beans
flour

Now try this!

- ## Draw two more things on each shelf.

Developing Numeracy
Measures, Shape and Space
Year 2
© A & C Black

Teachers' note Introduce a kilogram and a gram weight and give the children experience of holding a 1 kg weight alongside other objects. Young children often confuse size with weight; the objects above include ones which look large but are quite light. It would be useful to have many of these items in the classroom for children to test (mask the weights on the packets first).

Grams or kilograms?

- **You can use** grams **or** kilograms **to measure these things. Write which is best.**

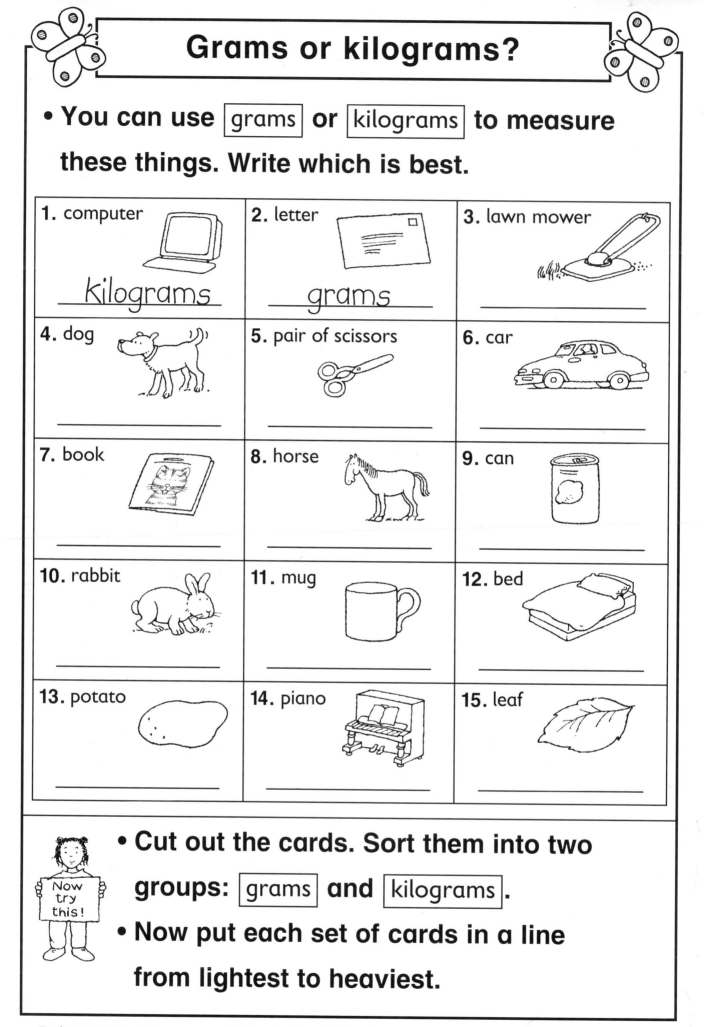

1. computer	2. letter	3. lawn mower
kilograms	_grams_	_____
4. dog	5. pair of scissors	6. car
_____	_____	_____
7. book	8. horse	9. can
_____	_____	_____
10. rabbit	11. mug	12. bed
_____	_____	_____
13. potato	14. piano	15. leaf
_____	_____	_____

Now try this!

- **Cut out the cards. Sort them into two groups:** grams **and** kilograms **.**
- **Now put each set of cards in a line from lightest to heaviest.**

Teachers' note The children can complete the extension activity in pairs to encourage discussion. Use a range of comparative vocabulary for this lesson, for example: Which weighs more/the least? Which is heavier/lighter? For the NTT, children can discuss and explain their order in the plenary session.

Developing Numeracy Measures, Shape and Space Year 2 © A & C Black

Guessing game

☆ Match an object in the room with an amount in the grid.

☆ If your partner agrees, place a counter on the circle.

☆ If your partner disagrees, write the name of the object and have another turn.

☆ Take turns to choose a different object.

☆ The winner is the first to get four counters in a line.

Play this game with a partner. You need counters in two colours.

Teachers' note Point out to the children that they are to estimate the weights of the objects. They may need to be given a kilogram weight to help them. As objects that weigh about a kilogram are harder to find, it would be helpful to have, for example, food and drink items around this weight in the classroom. During the plenary, discuss the objects that children have disagreed on during their game.

Developing Numeracy
Measures, Shape and Space
Year 2
© A & C Black

Weighty problems

- **Cut out the cards.**

- **Match the scales to the correct measurement.**

just over 1 kilogram	nearly 4 kilograms	about half a kilogram	just less than 6 kilograms
just under 1 kilogram	nearly 3 kilograms	about 5 kilograms	just less than 2 kilograms

Teachers' note As an extension, ask the children to order the scales from the lightest to heaviest. They could complete this extension in pairs to encourage discussion. Use a range of comparative vocabulary for this lesson, for example: Which weighs most/least? Which is heavier/lighter? Is it just more than or just less than?

Developing Numeracy Measures, Shape and Space Year 2 © A & C Black

Which holds more?

- **Colour the container you think holds** more .

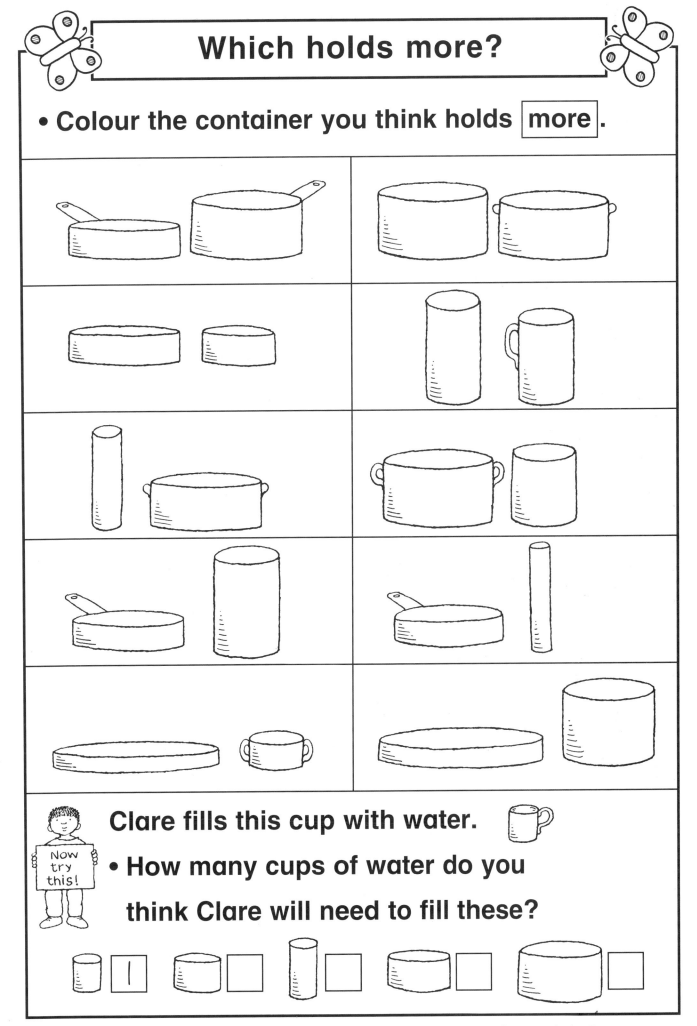

Clare fills this cup with water.

- **How many cups of water do you think Clare will need to fill these?**

Teachers' note Begin by showing the children a variety of litre containers of different shapes, for example tall and thin, short and wide, to help them make judgements about these pictures. Young children can often confuse height with how much something holds, e.g. thinking a tall thin container has a larger capacity than a short wide container. Discuss their decisions in the plenary session.

Developing Numeracy
Measures, Shape and Space
Year 2
© A & C Black

19

Bring and buy

- **Cut out the cards.** • **Sort them onto the correct table.**

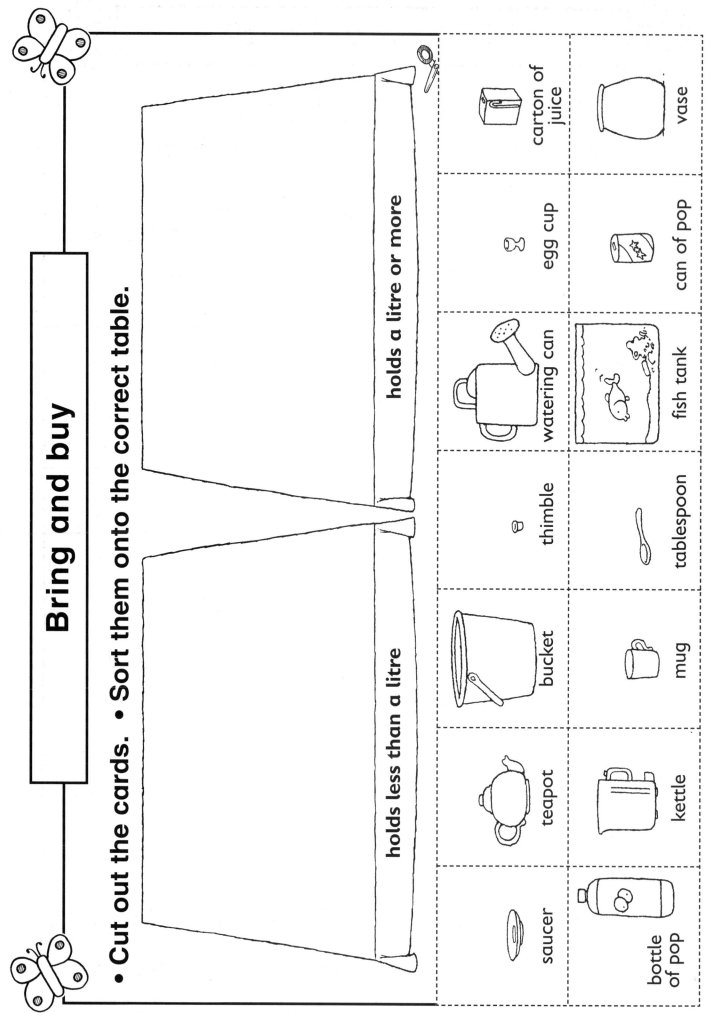

holds a litre or more

holds less than a litre

carton of juice	egg cup	watering can	thimble	bucket	teapot	saucer
vase	can of pop	fish tank	tablespoon	mug	kettle	bottle of pop

Teachers' note Show the children a variety of litre containers at the start of the lesson (see Teachers' note on page 19). If possible, have available some of the items shown on the cards for the children to test. As an extension, ask the children to draw two other items to go on each table.

Developing Numeracy
Measures, Shape and Space
Year 2
© A & C Black

Millilitres or litres?

- **You can use** millilitres **or** litres **to measure the capacity of these. Write which is best.**

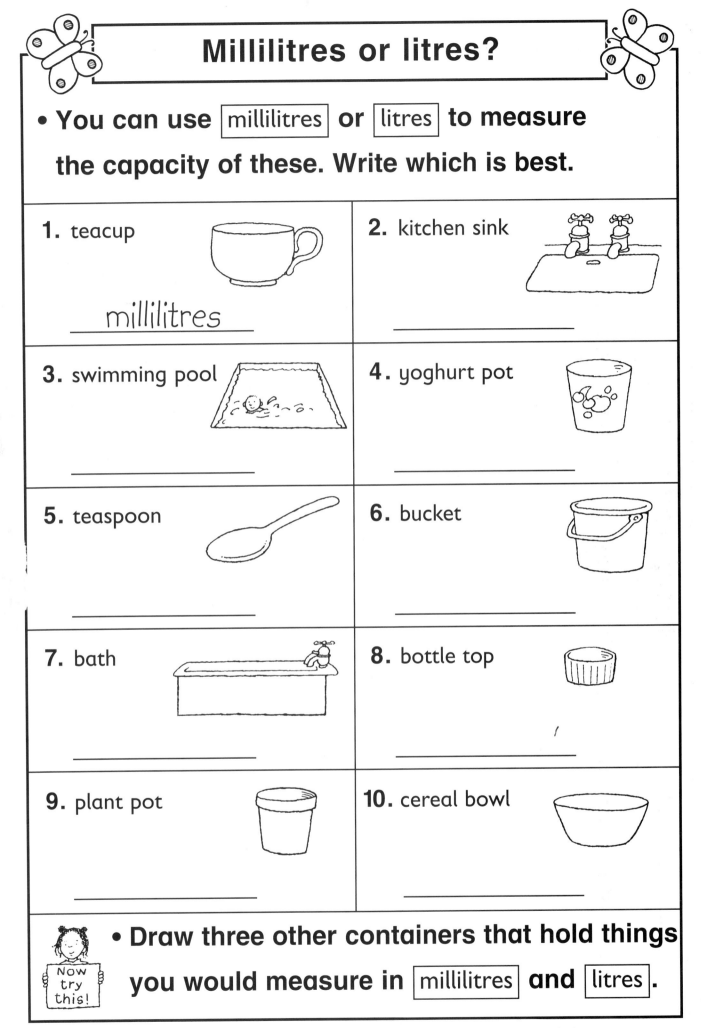

1. teacup

millilitres

2. kitchen sink

_____

3. swimming pool

_____

4. yoghurt pot

_____

5. teaspoon

_____

6. bucket

_____

7. bath

_____

8. bottle top

_____

9. plant pot

_____

10. cereal bowl

_____

Now try this!

- **Draw three other containers that hold things you would measure in** millilitres **and** litres **.**

Teachers' note The children can compare answers with a partner. It would be useful to have some of these items, or similar, in the classroom for the children to test using water or sand. Explain to the children that a millilitre is the amount of water that would fit into a centimetre cube and that a thousand of these make one litre.

Developing Numeracy
Measures, Shape and Space
Year 2
© A & C Black

Find it

• **Think of an object that holds each amount.**

Draw and name it.

1. less than a litre cat bowl _cat bowl_	**2.** about a litre _____
3. about half a litre _____	**4.** about two litres _____
5. more than five litres _____	**6.** more than ten litres _____
7. between three and five litres _____	**8.** about five litres _____

Now try this!

• **How many bottles of water will fill the bucket?** _____

half a litre   10 litres

Teachers' note Show the children a variety of litre containers at the start of the lesson to help them appreciate the size of a litre. Capacities of objects could be estimated or measured using sand or water. A list of items or a variety of objects could be used to help the children with this sheet.

Developing Numeracy
Measures, Shape and Space
Year 2
© A & C Black

Waterworld

- **Colour the jugs to show the correct amount of water.**

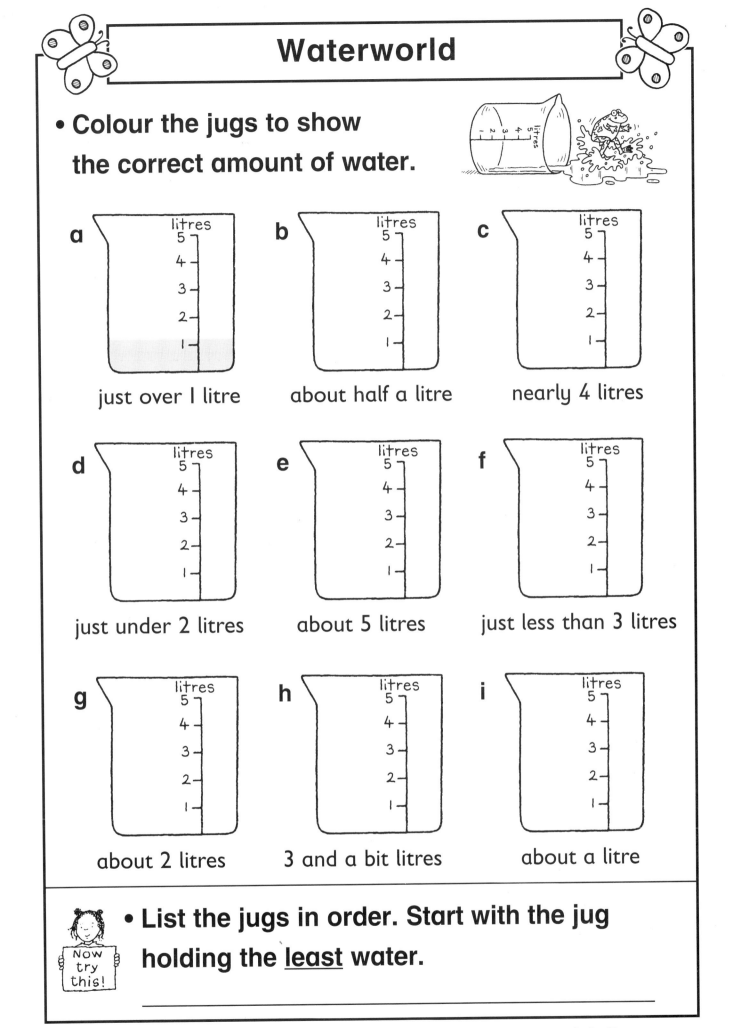

a just over 1 litre

b about half a litre

c nearly 4 litres

d just under 2 litres

e about 5 litres

f just less than 3 litres

g about 2 litres

h 3 and a bit litres

i about a litre

- **List the jugs in order. Start with the jug holding the least water.**

_____

Teachers' note Check that the children are confident with the vocabulary associated with capacity. Ensure that the children appreciate that a jug holding up to 5 litres is a very large container e.g. a bucket.

Developing Numeracy
Measures, Shape and Space
Year 2
© A & C Black

Gemma's calendar

- **Fill in the missing months in Gemma's calendar.**

Word-bank
January, December, August, April, September, July, October, May, November, February, June, March

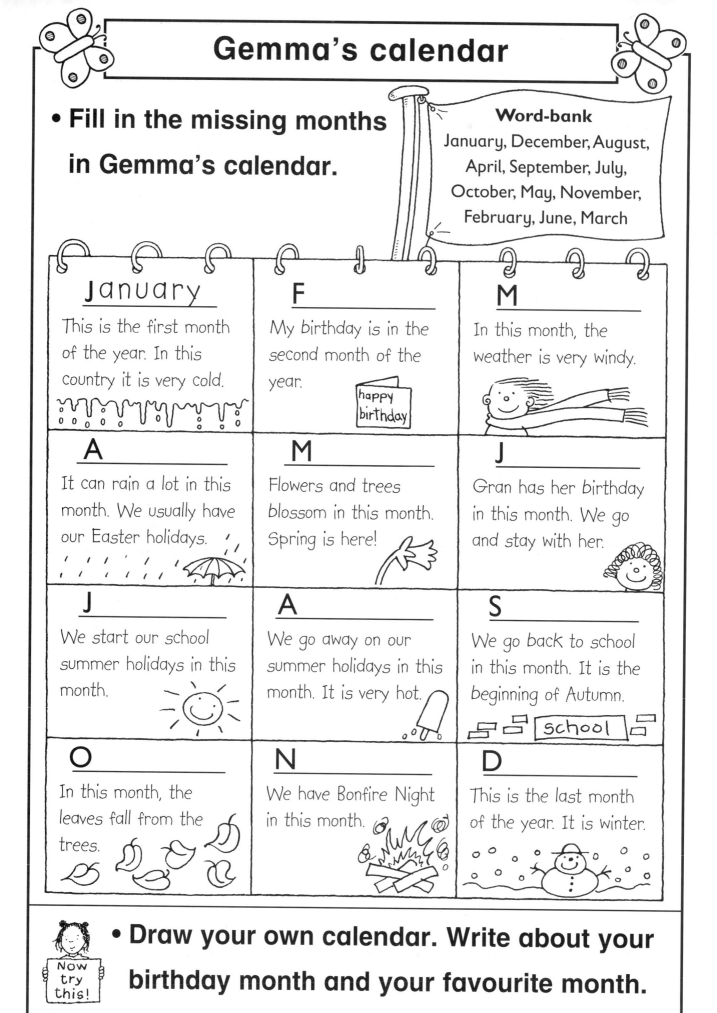

January This is the first month of the year. In this country it is very cold.	**F** My birthday is in the second month of the year.	**M** In this month, the weather is very windy.
A It can rain a lot in this month. We usually have our Easter holidays.	**M** Flowers and trees blossom in this month. Spring is here!	**J** Gran has her birthday in this month. We go and stay with her.
J We start our school summer holidays in this month.	**A** We go away on our summer holidays in this month. It is very hot.	**S** We go back to school in this month. It is the beginning of Autumn.
O In this month, the leaves fall from the trees.	**N** We have Bonfire Night in this month.	**D** This is the last month of the year. It is winter.

- **Draw your own calendar. Write about your birthday month and your favourite month.**

Now try this!

Teachers' note The initial letter of each month and the word-bank can be masked if desired. Read the calendar entries aloud, or invite a child to, and discuss experiences that the children have during the year.

**Developing Numeracy
Measures, Shape and Space
Year 2
© A & C Black**

24

Know your months

• **Colour the correct answer.**

1. The month that follows September is...

November   October

2. The month that follows April is...

June   May

3. The month that comes before July is...

May   June

4. The month that comes before January is...

February   December

5. Two months after May is...

June   July

6. Two months after October is...

September   December

7. Two months before March is...

January   May

8. Two months before November is...

September   January

• **Write all the months of the year. Start from July.**

Now try this!

Teachers' note Some children may need a list or wheel showing the months of the year to help them count forwards or backwards from a month.

Developing Numeracy
Measures, Shape and Space
Year 2
© A & C Black

How long?

• **How long do these activities take?**

Word-bank
second   minutes   hours
days   weeks

sleeping each night hours _____	blowing my nose _____	having lunch _____
school summer holidays _____	a school day _____	sneezing ahh _____ choo!
the weekend _____	writing my name Sarah _____	school assembly _____
having a bath _____	yawning yawn _____	a cartoon on TV _____

• **Write three activities that take <u>less than</u> a minute.**

NOW try this!

Teachers' note Discuss with the children that activities can be described in more than one unit of time, for example a maths lesson might be 1 hour or 60 minutes, or writing your name might take half a minute but could also be described in seconds. Encourage the children to consider a range of activities from their own experiences, for example book week, lunch hour.

Developing Numeracy
Measures, Shape and Space
Year 2
© A & C Black

What's the time, Mr Wolf?

- **Write these clock times in words.**

Word-bank

o'clock   half past
quarter past   quarter to

nine o'clock

- **Write each time an hour later.**

Teachers' note The hands on these clocks can be masked to provide a flexible resource.
Encourage the children to discuss what they might be doing at these times and to consider a
range of activities from their own experiences, for example the time school starts and ends.

Developing Numeracy
Measures, Shape and Space
Year 2
© A & C Black

27

Time match

These children are saying the times on the clocks.

• Match each child to the correct clock.

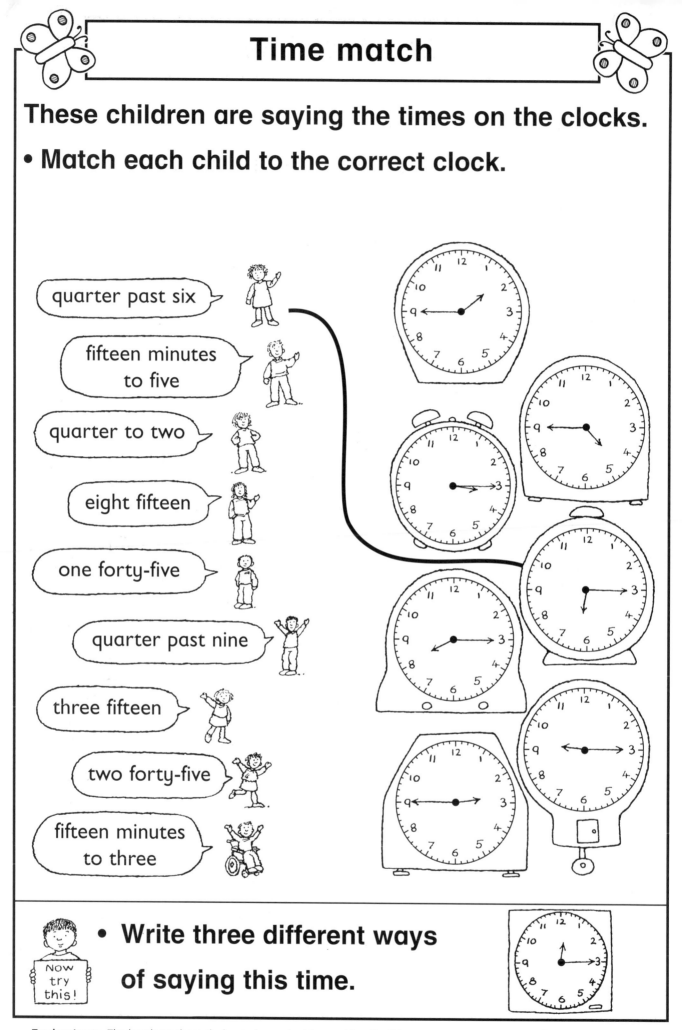

quarter past six

fifteen minutes to five

quarter to two

eight fifteen

one forty-five

quarter past nine

three fifteen

two forty-five

fifteen minutes to three

• **Write three different ways of saying this time.**

NOW try this!

Teachers' note The hands on these clocks can be masked to provide a flexible resource. Encourage the children to discuss what they might be doing at these times. Introduce the language of digital time prior to the activity, for example 'three fifteen' and warn children that some of the children are saying the same times in different ways.

Developing Numeracy Measures, Shape and Space Year 2 © A & C Black

Domino time

- **Cut out the dominoes.**

- **Join the clocks that show the same times.**

Teachers' note The times on the sheet can be masked and altered to provide extension or simplification. Some children may be able to tell the time to the nearest five minutes. The dominoes could be copied onto card and laminated for a more permanent resource.

Developing Numeracy
Measures, Shape and Space
Year 2
© A & C Black

Quiz time

☆ Cut out the cards. Give one to a partner.

☆ Your partner reads out the times in words on their card.

☆ For each time, **you** find a clock on **your** card that matches the 'letter' next to the clock.

☆ All correct answers win one point.

☆ When your partner has finished, read out the times on **your** card. Can your partner score more points than you?

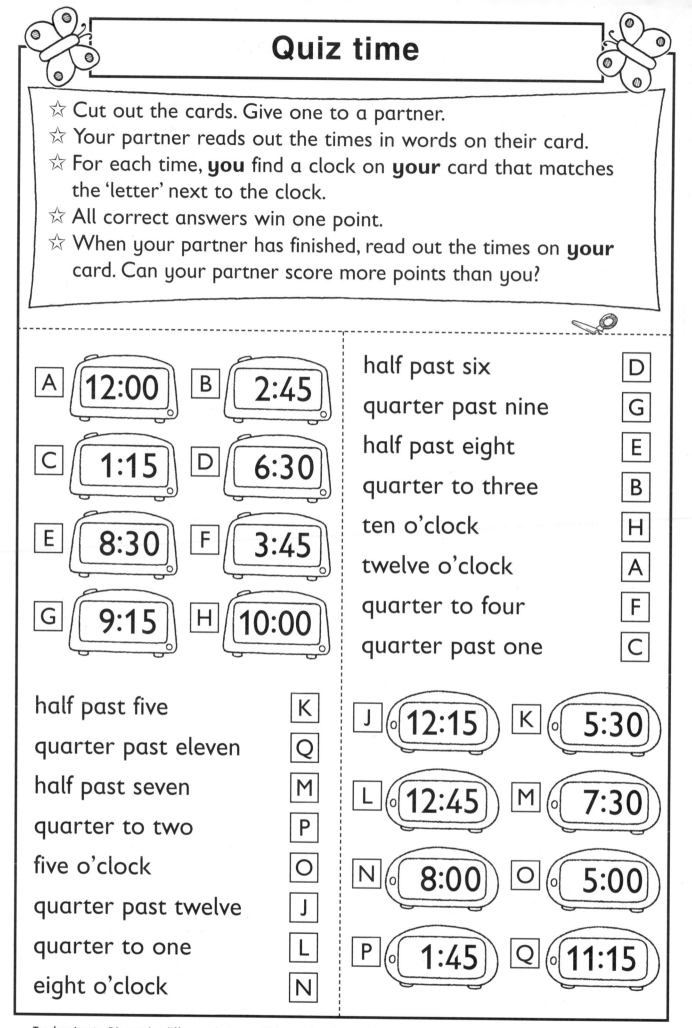

A 12:00	B 2:45
C 1:15	D 6:30
E 8:30	F 3:45
G 9:15	H 10:00

half past six   D

quarter past nine   G

half past eight   E

quarter to three   B

ten o'clock   H

twelve o'clock   A

quarter to four   F

quarter past one   C

half past five   K

quarter past eleven   Q

half past seven   M

quarter to two   P

five o'clock   O

quarter past twelve   J

quarter to one   L

eight o'clock   N

J 12:15	K 5:30
L 12:45	M 7:30
N 8:00	O 5:00
P 1:45	Q 11:15

Teachers' note Discuss the difference between digital and analogue clocks with the children. Read the instructions together and suggest that the children keep score by ticking correct answers. The times on the sheet can be masked and altered to provide extension or simplification. Some children may be able to tell the time to the nearest five minutes.

Developing Numeracy
Measures, Shape and Space
Year 2
© A & C Black

Shape collector

- ## Play this game with a partner.

  ☆ Take turns to move your counter to a **touching** square.

  ☆ Read the shape name. Collect the shape if you do not already have it.

  ☆ Colour the shape name on your score sheet.

  ☆ The first to collect all six shapes wins.

You will need a collection of solid shapes and a counter each.

Start Put your counters here	cube	sphere	cube	cuboid
cube	cylinder	sphere	pyramid	cube
sphere	pyramid	cube	cone	sphere
pyramid	cone	cuboid	cube	cylinder

Score sheet

Name _____

cube	sphere
cuboid	pyramid
cylinder	cone

Score sheet

Name _____

cube	sphere
cuboid	pyramid
cylinder	cone

Teachers' note Begin the lesson by reminding the children of the properties and names of the 3-D shapes listed. Each pair will need two of each shape listed, two counters and one copy of this sheet. Explain that at each turn they can only move across, up or down one square (not diagonally), and that they cannot move to a square that is already occupied.

Developing Numeracy
Measures, Shape and Space
Year 2
© A & C Black

31

Shape puzzle

- **Find a solid shape to match each picture. Place it on top of the picture.**

- **Join each shape to its name and description.**

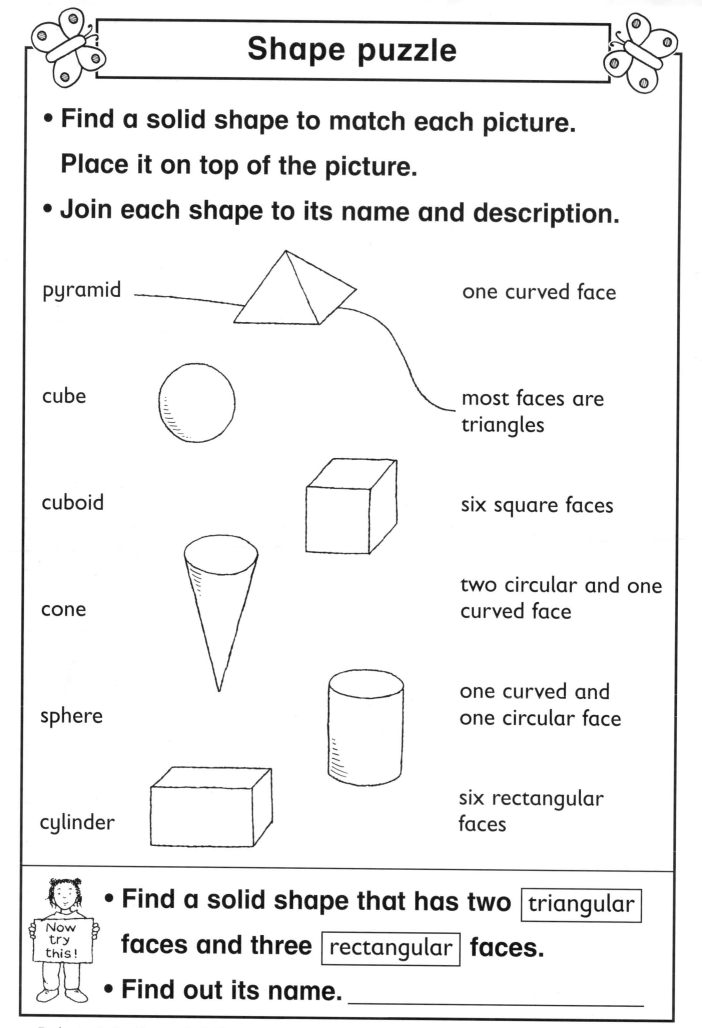

pyramid

cube

cuboid

cone

sphere

cylinder

one curved face

most faces are triangles

six square faces

two circular and one curved face

one curved and one circular face

six rectangular faces

Now try this!

- **Find a solid shape that has two** | triangular |

faces and three | rectangular | **faces.**

- **Find out its name.** _____

Teachers' note Provide a set of solid shapes. Children could work in pairs to discuss the shapes. Some children find recognising three dimensional shapes from pictures quite difficult. These children could be given the correct shapes and asked to match them to each picture, rather than choosing from a large set of shapes. For the extension activity, the children will need a triangular prism.

Developing Numeracy Measures, Shape and Space Year 2 © A & C Black

Feely bag fun

- **Guess which shape each child is describing.**

1.

My shape is like a ball. It has one curved face and no corners.

sphere

2.

My shape has two circular faces and one curved face. It reminds me of a can.

_____

3.

My shape is like a dice. It has six square faces and eight corners.

_____

4.

My shape has one circular face and one curved face. It has a point.

_____

5.

My shape is like a box. It has six rectangular faces.

_____

6.

My shape has one square face and four triangular faces. You see these shapes in Egypt.

_____

Now try this!

- **Hide a solid shape**
- **Describe it to a partner.**
- **Can your partner guess the shape?**

Teachers' note Encourage the children to play this game themselves. A feely bag can be used for a whole class version of this game. Encourage them to ask further questions about the shape, for example: 'Does it have any curved faces?' 'How many corners does it have?'

Developing Numeracy
Measures, Shape and Space
Year 2
© A & C Black

Shape hunt

● **Play this game with a partner.**

☆ You need a dice, two counters and some solid shapes.

☆ Take turns to roll the dice and move your counter.

☆ Find a shape that matches. If you cannot, go back three places.

☆ The winner is the first to reach 'finish'.

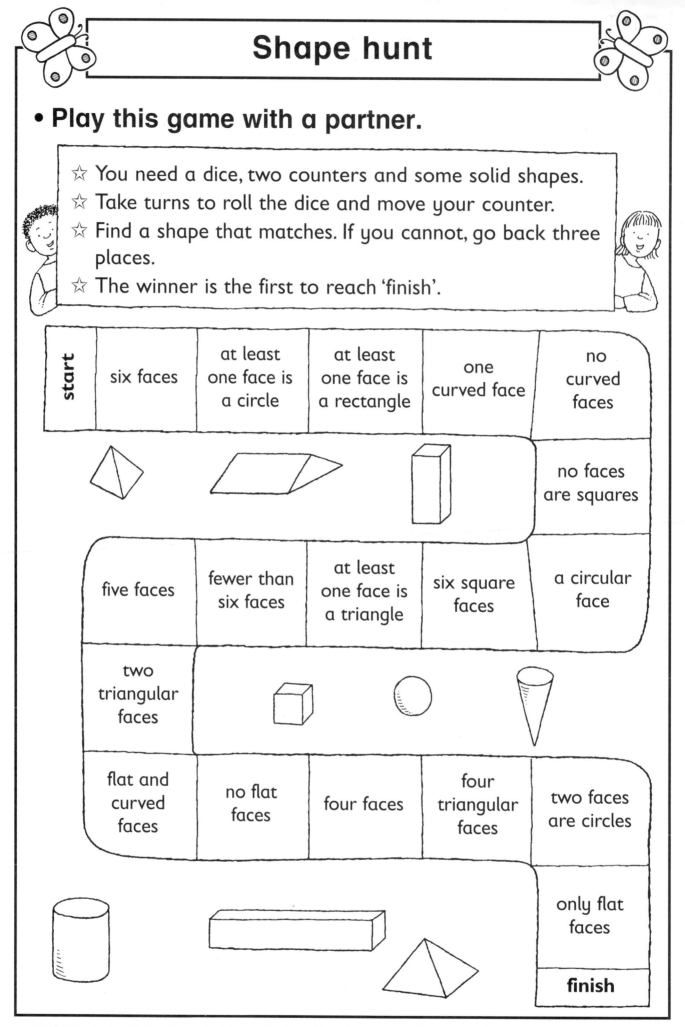

Developing Numeracy
Measures, Shape and Space
Year 2
© A & C Black

Skeletons

- **Build each skeleton shape.**

- **Count the number of** edges **and** corners .

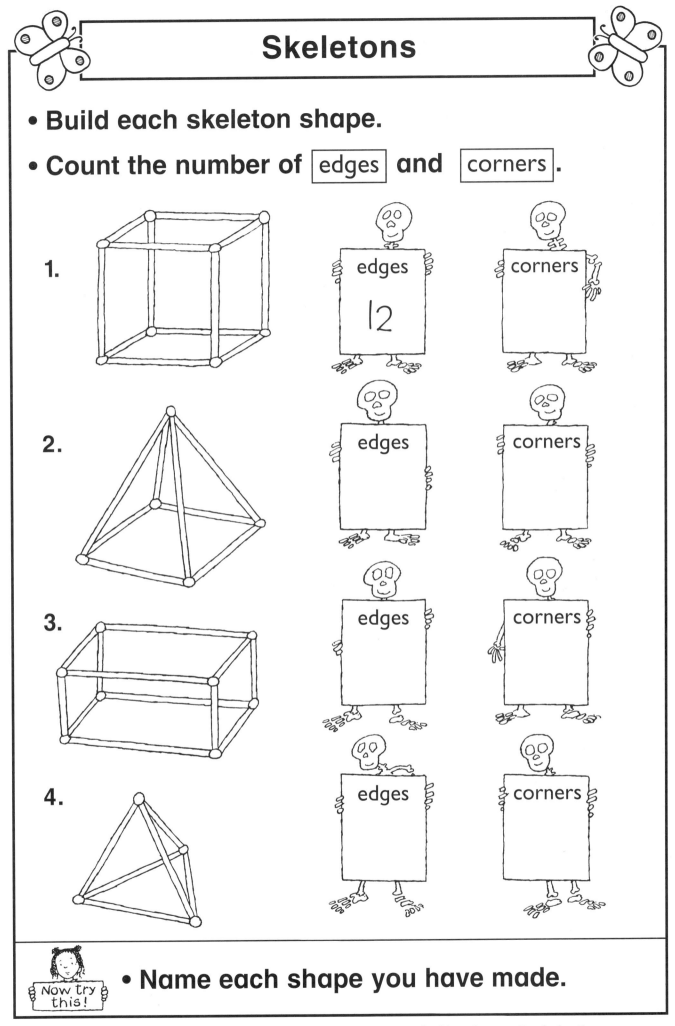

1. edges

   12

   corners

2. edges

   corners

3. edges

   corners

4. edges

   corners

Now try this!

- **Name each shape you have made.**

Teachers' note The children can build these shapes from materials in a construction kit, or from straws or cocktail sticks with plasticine.

Developing Numeracy
Measures, Shape and Space
Year 2
© A & C Black

True or false?

- **Make each model from solid shapes.**
- **Tick** `true` **or** `false` **for each statement.** ✔

1. This is made from four cubes.

✔ true ☐ false

2. This is made from a cube and a cone.

☐ true ☐ false

3. This is made from a pyramid and a cuboid.

☐ true ☐ false

4. This is made from a sphere and two cylinders.

☐ true ☐ false

5. This is made from two cubes and a cuboid.

☐ true ☐ false

6. This is made from a cone and a cube.

☐ true ☐ false

7. This is made from a cylinder and two cubes.

☐ true ☐ false

8. This is made from a sphere and a pyramid.

☐ true ☐ false

Now try this!

- **Use four interlocking cubes to make as many different models as you can.**

Teachers' note The children should use solid shapes and interlocking cubes for this activity. For the statements that are not true, ask the children to suggest what the correct statements might be. Encourage the children to compare shapes with a partner.

**Developing Numeracy
Measures, Shape and Space
Year 2
© A & C Black**

Shape stories

• Underline the shape in each story and draw it.

The fairies made a <u>circle</u> around the sad gnome.

The police ran into the town square and saw the robber.

Jack was happy to be in the school band. He played his triangle and smiled.

The big wheel looked like a huge sparkling octagon.

"My watch was the shape of a hexagon," Jane told the police.

Max saw the mad professor draw a large pentagon in the sand.

• Write a 'rectangle' shape story.

Now try this!

Teachers' note The children can draw a range of different shapes for each story, for example triangles in different orientations and sizes. Encourage the children to look out for uses of shape words in their own story books.

Developing Numeracy
Measures, Shape and Space
Year 2
© A & C Black

Wrapping paper

This wrapping paper shows different flat shapes.

1. Colour all the:

How many are there?

triangles yellow △ 8

squares blue □ —

pentagons pink ⬠ —

hexagons green ⬡ —

octagons orange ⯃ —

circles purple ○ —

2. What is the name of the shape that is

not coloured? _____

Teachers' note Discuss the properties of the 2-D shapes before beginning this activity. Ensure that the children appreciate that, for example, any five-sided shape is called a pentagon, not just the regular forms. Ask the children to create their own simple shape wrapping paper, allowing them a certain number of sticky paper shapes.

Developing Numeracy
Measures, Shape and Space
Year 2
© A & C Black

Shape cards

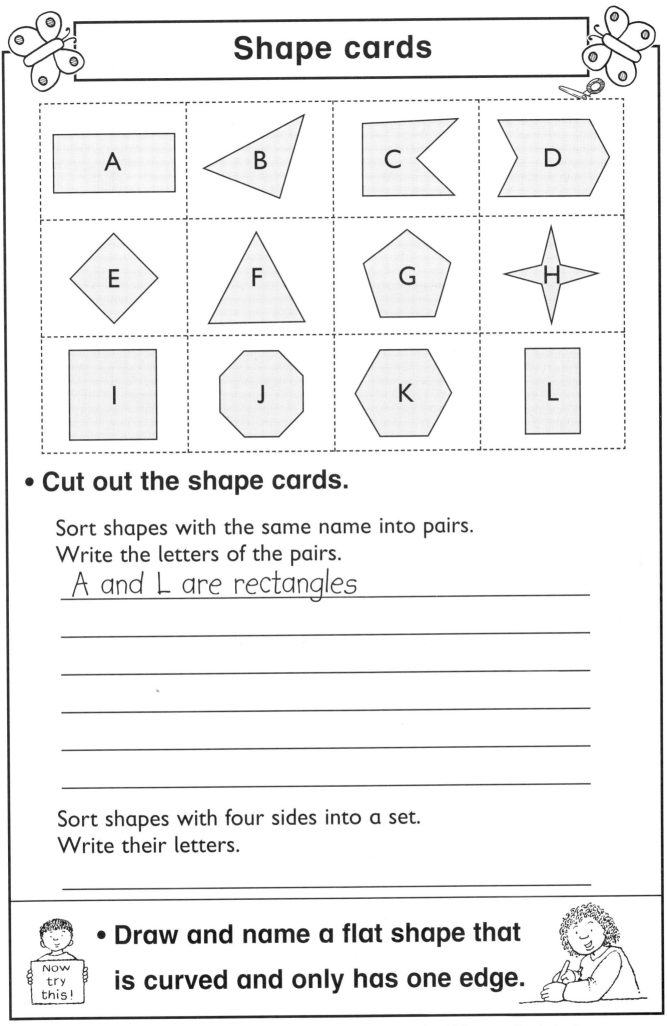

• **Cut out the shape cards.**

Sort shapes with the same name into pairs.
Write the letters of the pairs.

A and L are rectangles

Sort shapes with four sides into a set.
Write their letters.

• **Draw and name a flat shape that is curved and only has one edge.**

Now try this!

Teachers' note Some of the shapes on this sheet are regular and some are irregular. This is to ensure that the children do not develop the concept that only regular shapes are called hexagons, octagons, etc. Avoid using the word 'regular' when describing rectangles as, mathematically, 'regular' means that all the sides and all the angles of the shape are equal. NB The children do not need to be introduced to the words regular and irregular at this stage.

Developing Numeracy
Measures, Shape and Space
Year 2
© A & C Black

Picture perfect

This picture has been drawn using:

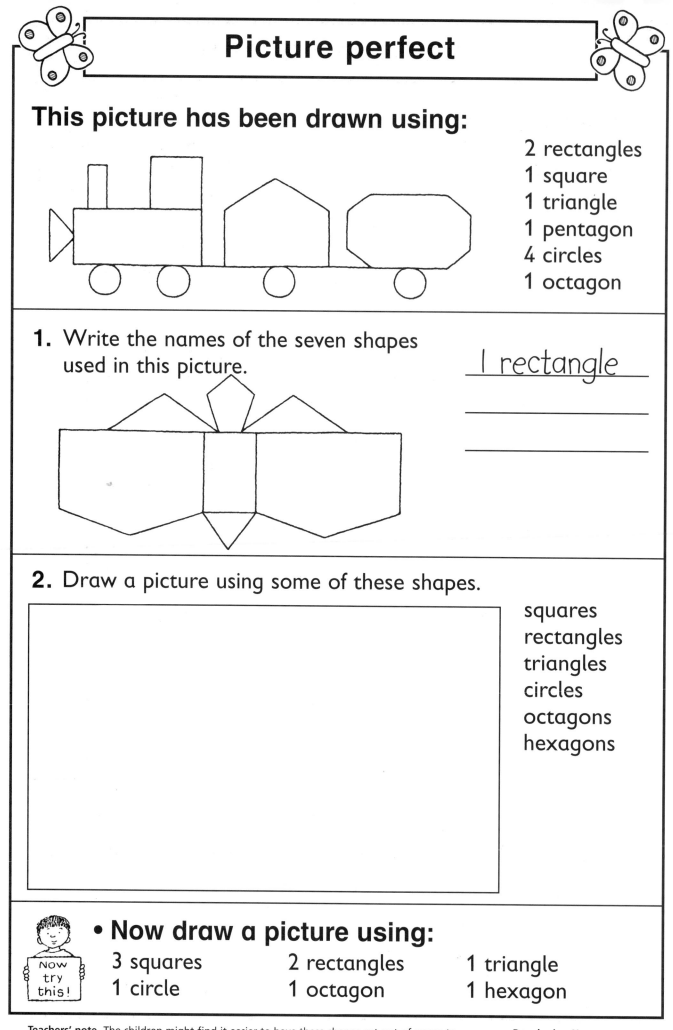

2 rectangles
1 square
1 triangle
1 pentagon
4 circles
1 octagon

1. Write the names of the seven shapes used in this picture.

<u>I rectangle</u>

_____

_____

2. Draw a picture using some of these shapes.

squares
rectangles
triangles
circles
octagons
hexagons

• **Now draw a picture using:**

NOW try this!

3 squares   2 rectangles   1 triangle
1 circle   1 octagon   1 hexagon

Teachers' note The children might find it easier to have these shapes cut out of paper to rearrange, or to use sticky paper shapes. The children could make larger pictures in this way and these could be labelled and displayed on the wall.

Developing Numeracy
Measures, Shape and Space
Year 2
© A & C Black

Pentagon patterns

- **Draw different pentagons on this dotty grid.**

The corners of the shapes must touch the dots.

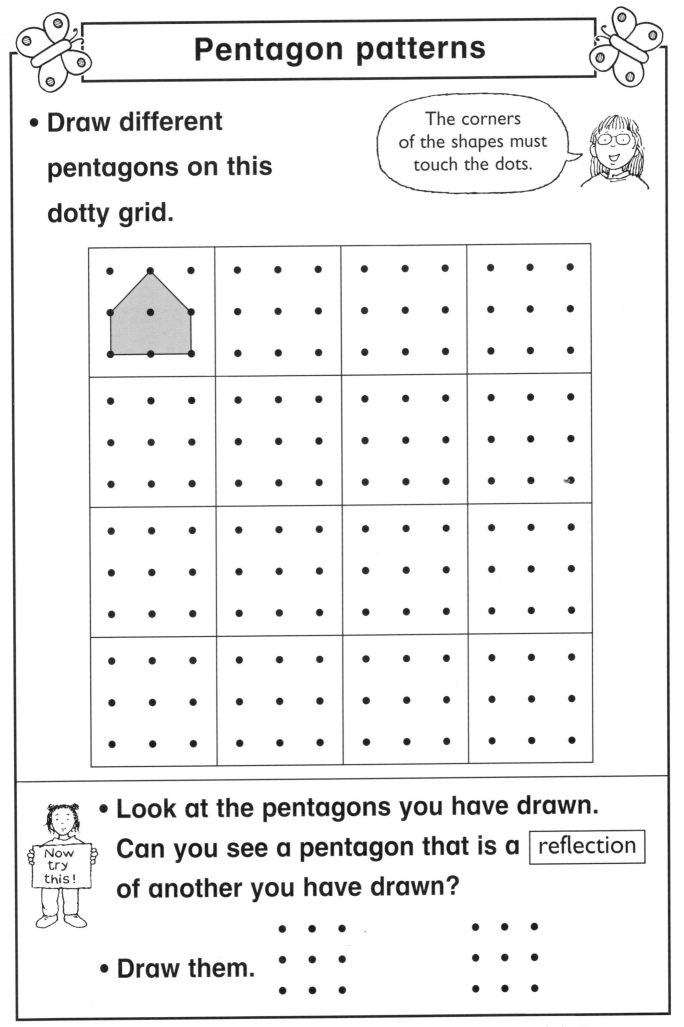

- **Look at the pentagons you have drawn.**

  Can you see a pentagon that is a reflection **of another you have drawn?**

- **Draw them.**

Teachers' note The children can work in pairs to discuss the extension activity. Ensure that they appreciate that pentagons are shapes with five straight sides, that do not have to be regular. Discuss the shapes drawn and look for reflections and rotations, cutting the shapes out and sorting if appropriate.

Developing Numeracy
Measures, Shape and Space
Year 2
© A & C Black

Tile teasers: 1

These shapes are made from tiles which touch along <u>at least</u> one side.

• Write the name of each shape.

Word-bank

rectangle
hexagon
octagon
decagon

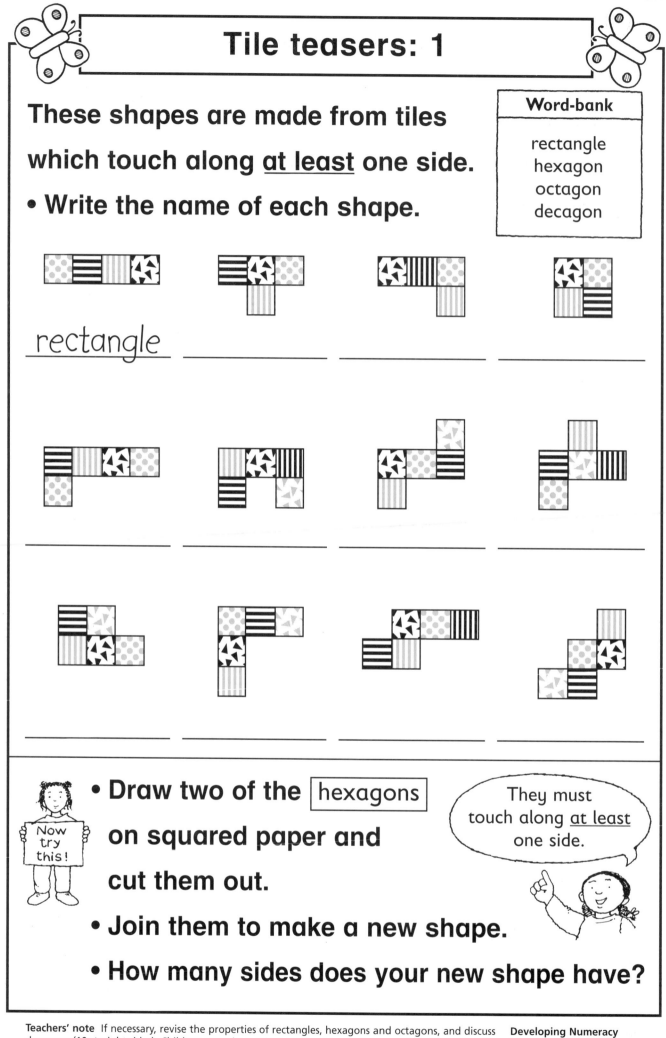

rectangle

_____

_____

_____

_____   _____   _____   _____

_____   _____   _____   _____

• Draw two of the ⬚hexagons⬚ on squared paper and cut them out.

They must touch along <u>at least</u> one side.

Now try this!

• Join them to make a new shape.

• How many sides does your new shape have?

Teachers' note If necessary, revise the properties of rectangles, hexagons and octagons, and discuss decagons (10 straight sides). Children sometimes miscount the sides of shapes such as these. Provide them with a coloured pencil and ask them to draw a line along each side as they count it. As a further extension, the children could explore hexominoes (six squares) in the same way.

**Developing Numeracy
Measures, Shape and Space
Year 2
© A & C Black**

Tile teasers: 2

These shapes are made from tiles which touch along <u>at least</u> one side.

• Write the name of each shape.

Word-bank

triangle
pentagon
hexagon
octagon

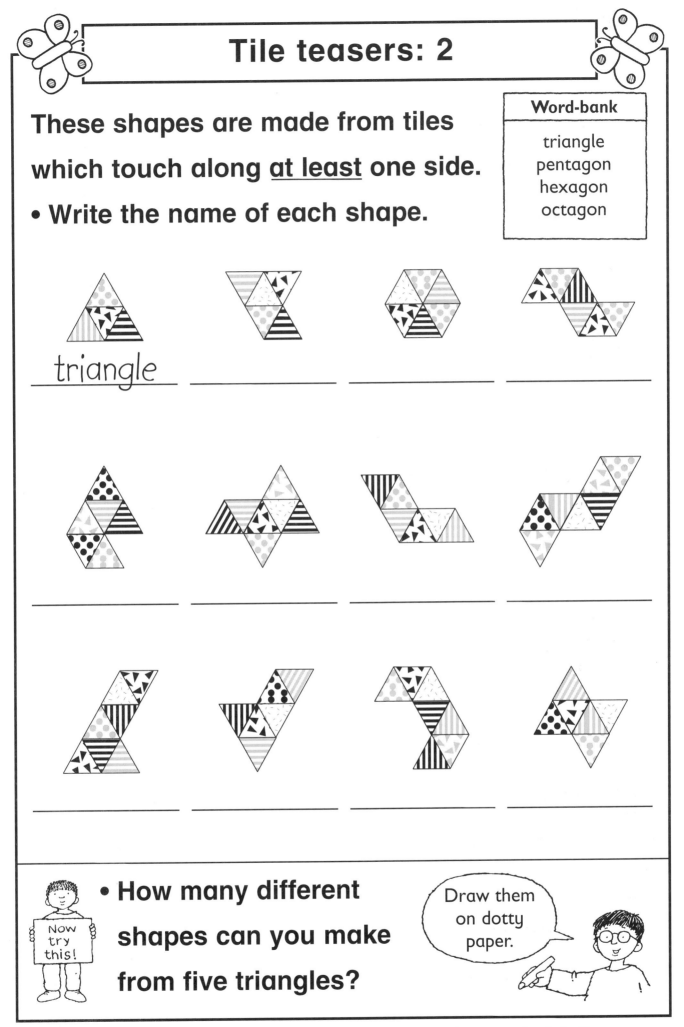

triangle _____ _____ _____

_____ _____ _____ _____

_____ _____ _____ _____

• **How many different shapes can you make from five triangles?**

Now try this!

Draw them on dotty paper.

Teachers' note If necessary, revise the properties of triangles, pentagons, hexagons and octagons. Children sometimes miscount the sides of shapes such as these. Provide them with a coloured pencil and ask them to draw a line along each side as they count it. The children will need dotty triangular (isometric) paper for the extension activity.

**Developing Numeracy
Measures, Shape and Space
Year 2**
© A & C Black

43

Animal symmetry

• **Play this game with a partner.**

☆ Cut out the cards, jumble them up and spread them
face down.

☆ Take turns to pick two cards. If they are **reflections** keep
them. If not, put them back.

☆ The winner is the player with most cards.

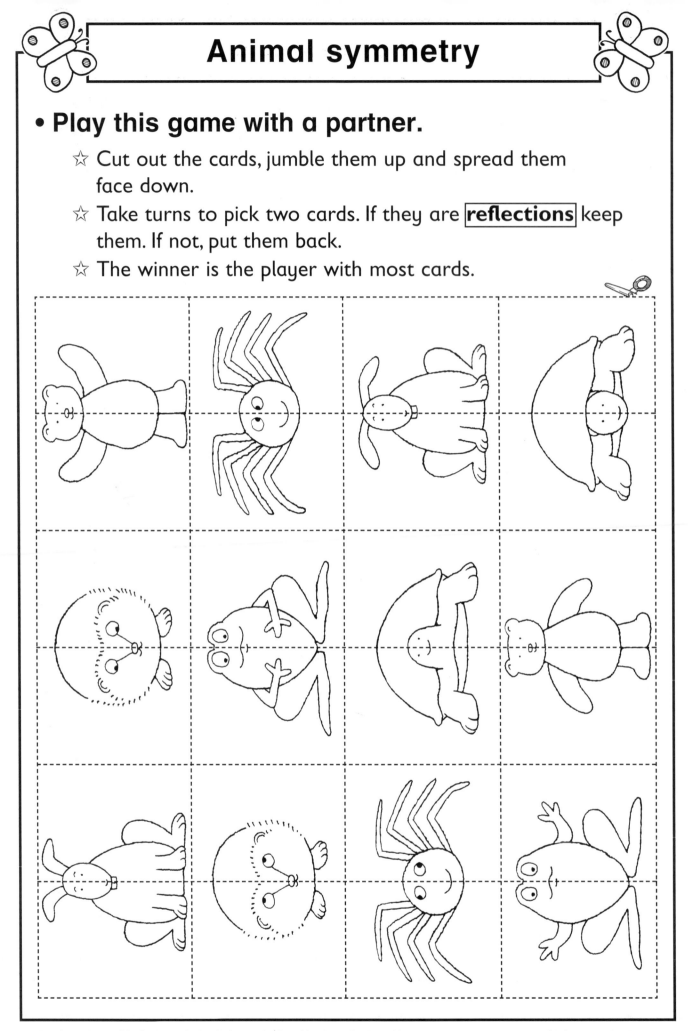

Teachers' note This sheet can be copied onto card and laminated to provide a more permanent resource. A variety of Pairs or Snap games can be played with them or, alternatively, the children can sort them independently.

**Developing Numeracy
Measures, Shape and Space
Year 2
© A & C Black**

Paper shapes

Mira has made these ⟨symmetrical⟩ shapes from paper.

- Draw a line to show where she folded them.

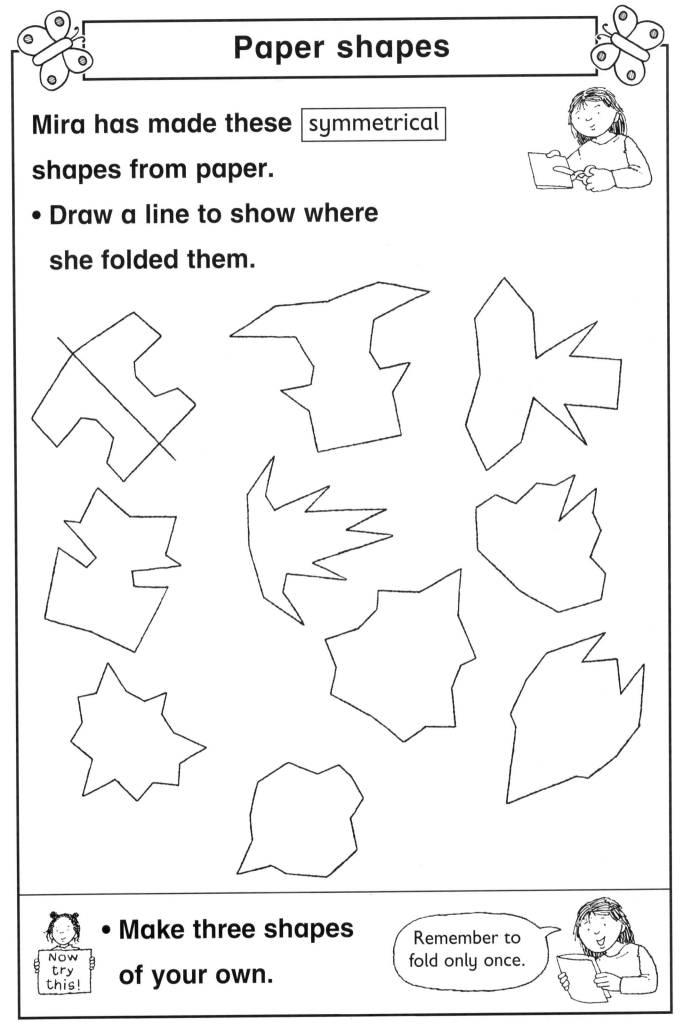

Teachers' note Start by modelling how to make symmetrical shapes by folding and cutting paper. If appropriate, provide the children with small mirrors for checking their solutions. Introduce the terms 'line of symmetry' or 'mirror line' instead of 'fold line', once the children begin to understand the idea of symmetry.

**Developing Numeracy
Measures, Shape and Space
Year 2**
© A & C Black

45

Spaceman Sam

Sam the Spaceman visits two planets. On one planet **everything** is symmetrical . On the other planet **nothing** is symmetrical.

• Join each object to the correct planet.

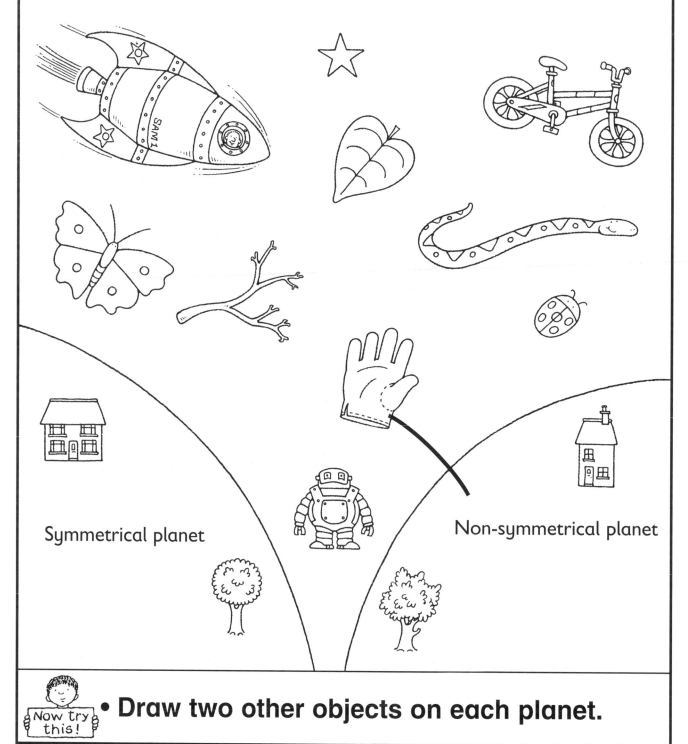

Symmetrical planet

Non-symmetrical planet

• **Draw two other objects on each planet.**

Now try this!

Teachers' note Some children may be able to complete this activity without a mirror. Ensure that the children realise that lines of symmetry are not only horizontal or vertical, for example the leaf. If able, children could draw the lines of symmetry onto each of the shapes on the symmetrical planet.

Developing Numeracy
Measures, Shape and Space
Year 2
© A & C Black

Road signs

- **Draw a** line of symmetry **on each road sign.**

- **Draw three symmetrical signs of your own for the classroom.**
- **Mark a line of symmetry.**

Teachers' note Ensure that the children realise that it is the symbol on the sign that they are looking at, not the sign or line shape of the sign itself, for example a circle has an infinite number of lines of symmetry, but the 30 speed limit sign has only one. Some children could be asked to find all the lines of symmetry for each road sign.

Developing Numeracy
Measures, Shape and Space
Year 2
© A & C Black

47

Dotty's friends

Dotty the ladybird has lots of friends.

• Colour the spots so that each of Dotty's friends has a symmetrical pattern.

The first one has been done for you.

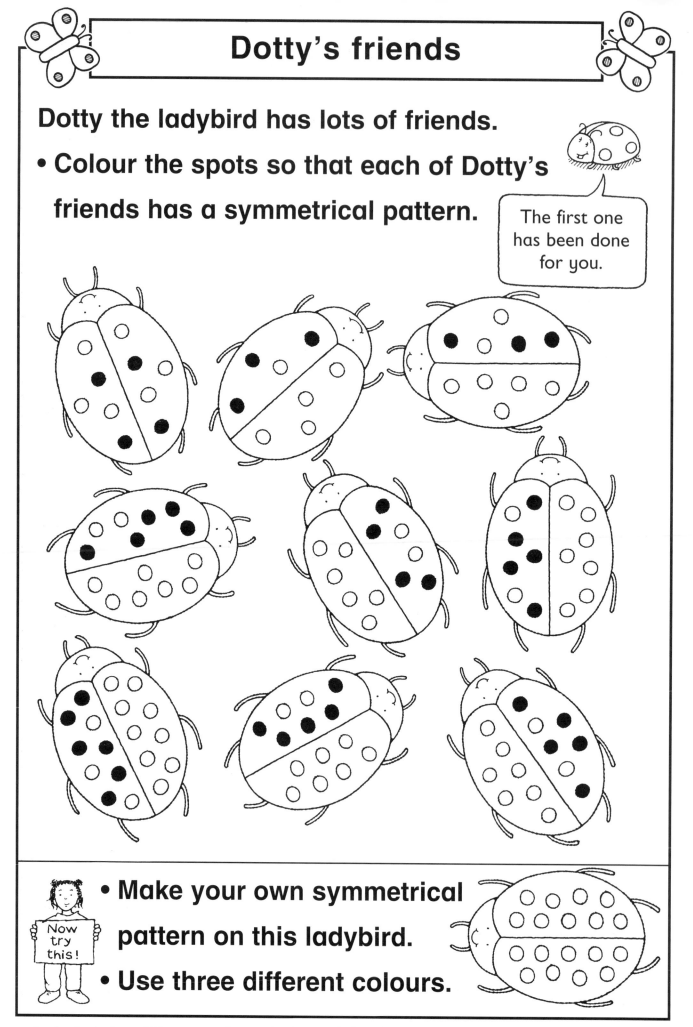

• Make your own symmetrical pattern on this ladybird.

• Use three different colours.

Now try this!

Teachers' note Some children may need to turn the sheets so the line of symmetry is vertical each time. As a further activity, the children could make large symmetrical ladybirds for display, using small circles of different-coloured sticky paper. Some ladybirds could have more than one line of symmetry.

Developing Numeracy
Measures, Shape and Space
Year 2
© A & C Black

Kite flying

- **You need three different coloured pencils.**

- **Colour each kite using all three colours.**

- **Make each kite symmetrical and each kite different.**

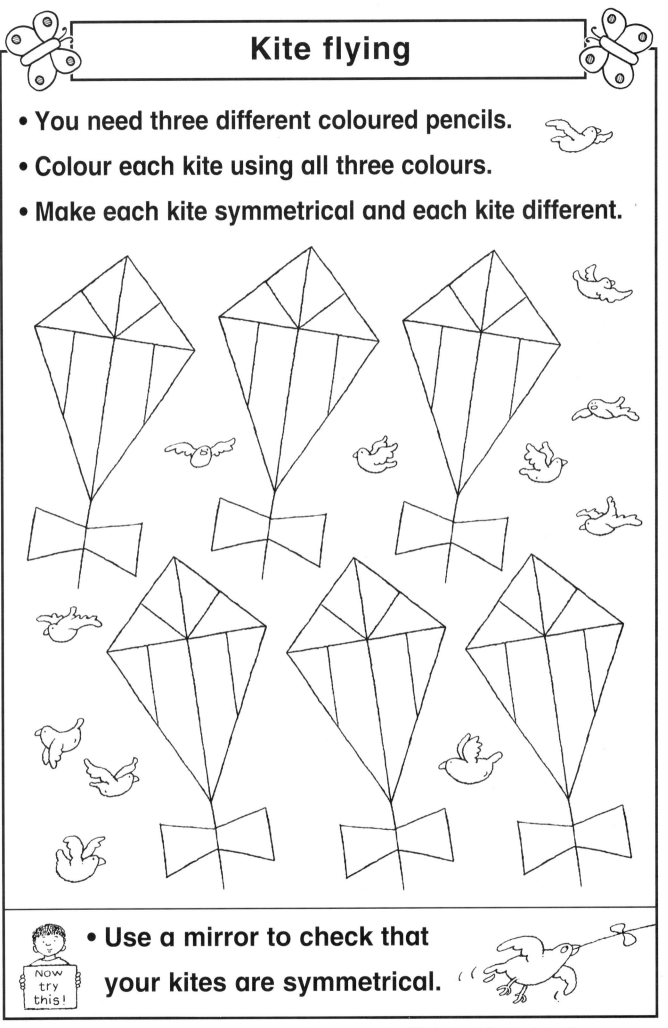

- **Use a mirror to check that your kites are symmetrical.**

Now try this!

Teachers' note Children often find using a mirror to check for symmetry quite difficult. Some fail to realise that the mirror must be lifted to see that the pattern underneath matches what can be seen in the mirror. Other perspex equipment is available that allows children to see the pattern through the perspex at the same time as seeing the reflection on the perspex.

Developing Numeracy
Measures, Shape and Space
Year 2
© A & C Black

Fruit shop

- **Look at the fruit on the shelf.**
- **Complete each sentence using any of these words or phrases.**

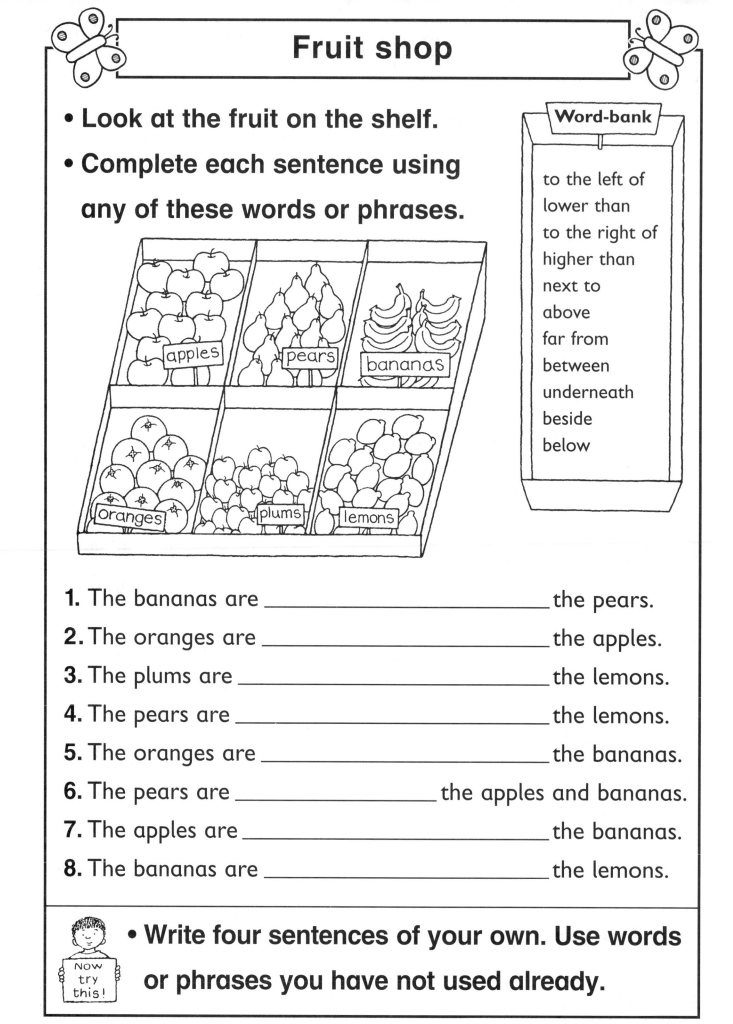

Word-bank

to the left of
lower than
to the right of
higher than
next to
above
far from
between
underneath
beside
below

1. The bananas are _____ the pears.

2. The oranges are _____ the apples.

3. The plums are _____ the lemons.

4. The pears are _____ the lemons.

5. The oranges are _____ the bananas.

6. The pears are _____ the apples and bananas.

7. The apples are _____ the bananas.

8. The bananas are _____ the lemons.

Now try this!

- **Write four sentences of your own. Use words or phrases you have not used already.**

Teachers' note The children can be asked further questions about the positions of the fruit. Encourage them to see opposites, for example that if the bananas are to the right of the pears, then the pears are to the left of the bananas. The children can make a collection of position words for display.

**Developing Numeracy
Measures, Shape and Space
Year 2**
© A & C Black

Toy shop

- **Tick the statements that are** $\boxed{\text{true}}$ **.**   ☑

1. The doll is between the robot and the ball. ✔

2. The ball is to the right of the doll. ☐

3. The dinosaur is beside the car. ☐

4. The teddy is underneath the ball. ☐

5. The doll is higher than the car. ☐

6. The ball is above the dinosaur. ☐

7. The robot is next to the ball. ☐

8. The teddy is above the robot. ☐

Now try this!

- **Write four sentences of your own about the position of the toys.**

Teachers' note The children can be asked further questions about the positions of the toys. Encourage them to see opposites, for example that if the doll is to the right of the robot, then the robot is to the left of the doll. The children can collect a list of position words for display.

Developing Numeracy
Measures, Shape and Space
Year 2
© A & C Black

Which letter?

- **Put your pencil on the dot. Follow the instructions.**

- **Which letter have you written?**

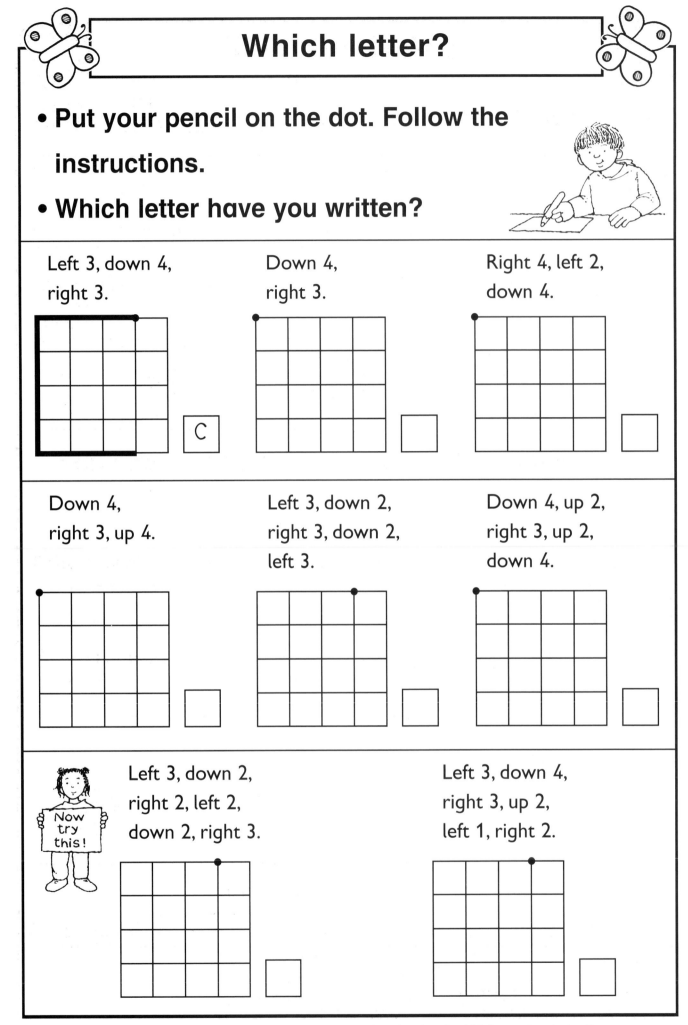

Left 3, down 4, right 3.

Down 4, right 3.

Right 4, left 2, down 4.

Down 4, right 3, up 4.

Left 3, down 2, right 3, down 2, left 3.

Down 4, up 2, right 3, up 2, down 4.

Now try this!

Left 3, down 2, right 2, left 2, down 2, right 3.

Left 3, down 4, right 3, up 2, left 1, right 2.

Teachers' note If necessary, revise 'left' and 'right'. Stress to the children that they should follow along the lines of the squares when completing this activity. As a further extension, the children can write instructions for other letters such as F, I, J, O, W. Instructions can be swapped with a partner and worked out using squared paper.

Developing Numeracy
Measures, Shape and Space
Year 2
© A & C Black

Safari survival!

Here is a plan of a safari park.

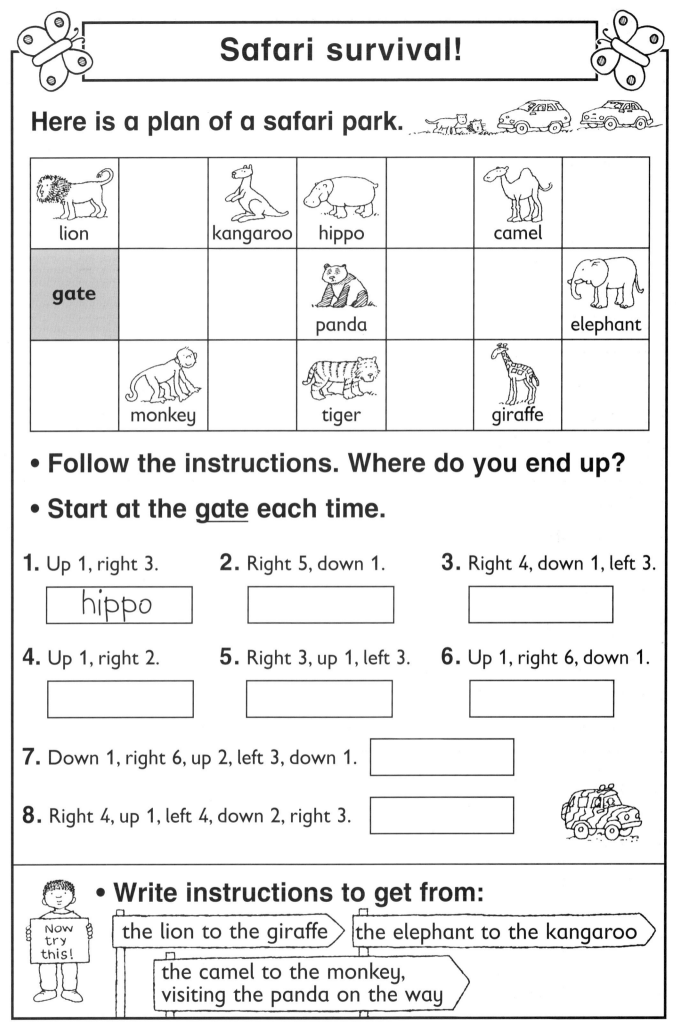

lion		kangaroo	hippo		camel	
gate			panda			elephant
	monkey		tiger		giraffe	

- **Follow the instructions. Where do you end up?**

- **Start at the gate each time.**

1. Up 1, right 3.

　hippo

2. Right 5, down 1.

3. Right 4, down 1, left 3.

4. Up 1, right 2.

5. Right 3, up 1, left 3.

6. Up 1, right 6, down 1.

7. Down 1, right 6, up 2, left 3, down 1.

8. Right 4, up 1, left 4, down 2, right 3.

- **Write instructions to get from:**

Now try this!

the lion to the giraffe

the elephant to the kangaroo

the camel to the monkey, visiting the panda on the way

Teachers' note This activity can be introduced during the first part of the lesson. Ask the children to listen to routes you describe or to make up their own. For children who experience difficulty with left and right, ask them to make an 'L' shape with their left fingers and thumb and to think of the 'L' as standing for 'Left'. Discuss the different possible answers to the extension activities.

Developing Numeracy
Measures, Shape and Space
Year 2
© A & C Black

53

Cat and mouse

The cat must try to catch the mouse!

You need:
- two counters
- a coin
- the four cards (face up)

Heads = move 2
Tails = move 1

☆ Decide who is 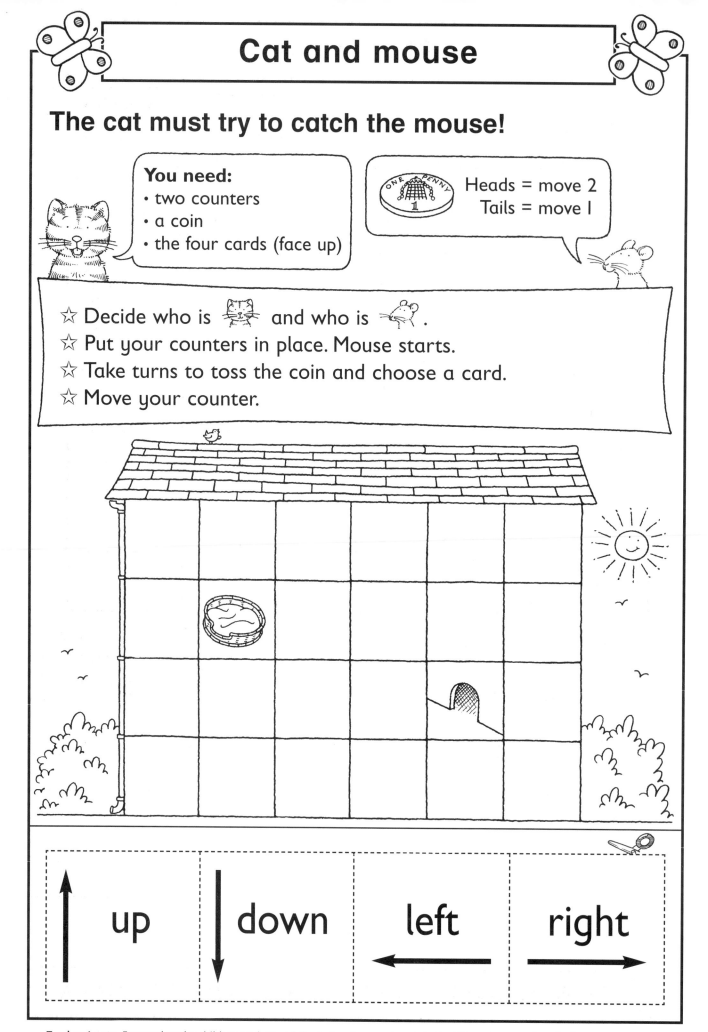 and who is .

☆ Put your counters in place. Mouse starts.

☆ Take turns to toss the coin and choose a card.

☆ Move your counter.

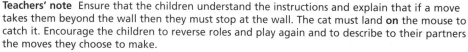

↑ up | ↓ down | left ← | right →

Teachers' note Ensure that the children understand the instructions and explain that if a move takes them beyond the wall then they must stop at the wall. The cat must land **on** the mouse to catch it. Encourage the children to reverse roles and play again and to describe to their partners the moves they choose to make.

Developing Numeracy
Measures, Shape and Space
Year 2
© A & C Black

Star maze

- **Place a counter on star 1.**

- **Move along the dotted line.**

- **Describe the route to a partner.**

Use these words.

Word-bank

left ←
right →
up ↑
down ↓

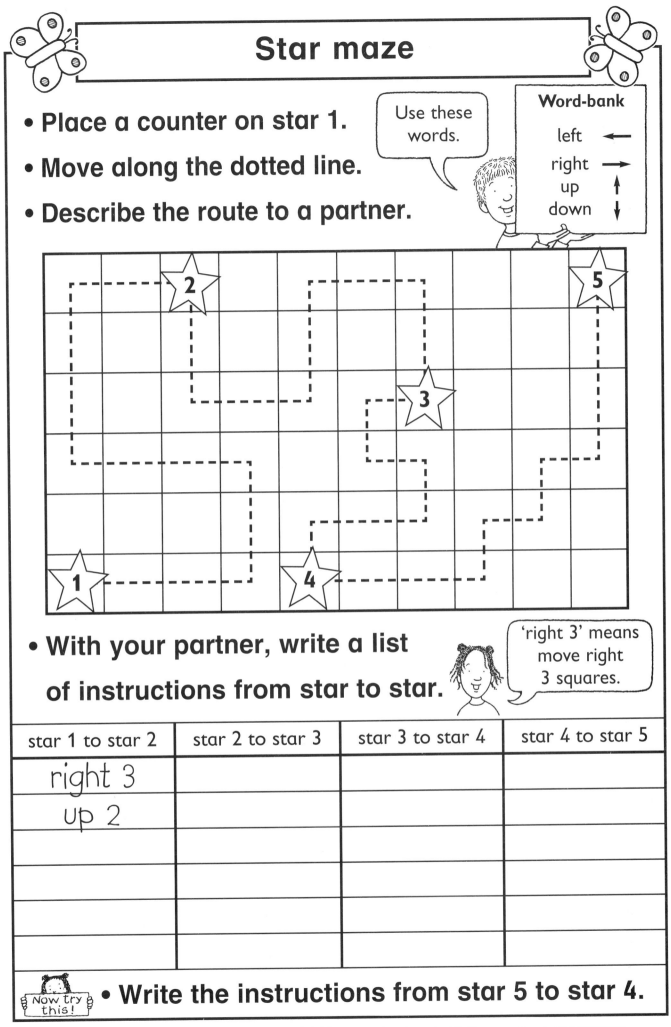

- **With your partner, write a list of instructions from star to star.**

'right 3' means move right 3 squares.

star 1 to star 2	star 2 to star 3	star 3 to star 4	star 4 to star 5
right 3			
up 2			

Now try this! • **Write the instructions from star 5 to star 4.**

Teachers' note As a further extension, give each pair of children a set of cards marked 'left', 'right', 'up' and 'down', and a counter each. The children should take turns to pick a card and, where possible, move in that direction until a new junction is reached. A target square can be agreed in advance and the winner is the first to reach it.

Developing Numeracy
Measures, Shape and Space
Year 2
© A & C Black

Robot rotations

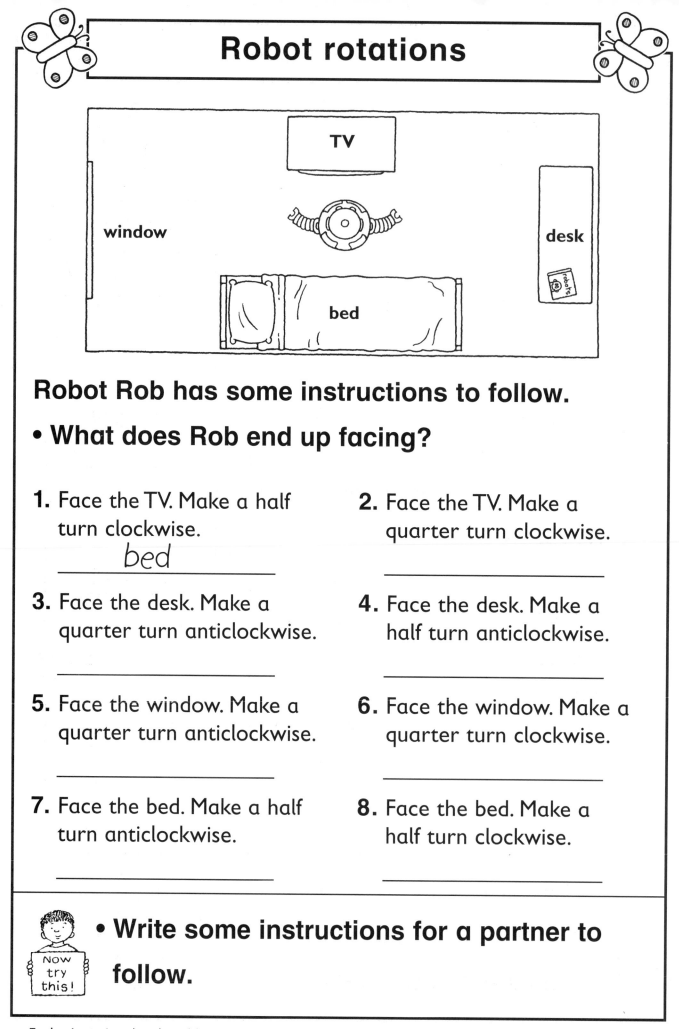

Robot Rob has some instructions to follow.

• What does Rob end up facing?

1. Face the TV. Make a half turn clockwise.

bed
_____

2. Face the TV. Make a quarter turn clockwise.

_____

3. Face the desk. Make a quarter turn anticlockwise.

_____

4. Face the desk. Make a half turn anticlockwise.

_____

5. Face the window. Make a quarter turn anticlockwise.

_____

6. Face the window. Make a quarter turn clockwise.

_____

7. Face the bed. Make a half turn anticlockwise.

_____

8. Face the bed. Make a half turn clockwise.

_____

• **Write some instructions for a partner to follow.**

Teachers' note Introduce the activity practically by asking a child to make different clockwise and anticlockwise turns. Explain to the children that the picture shows a room viewed from above. Encourage them to notice that the same result comes from a clockwise or anticlockwise half turn.

Developing Numeracy
Measures, Shape and Space
Year 2
© A & C Black

On the spot

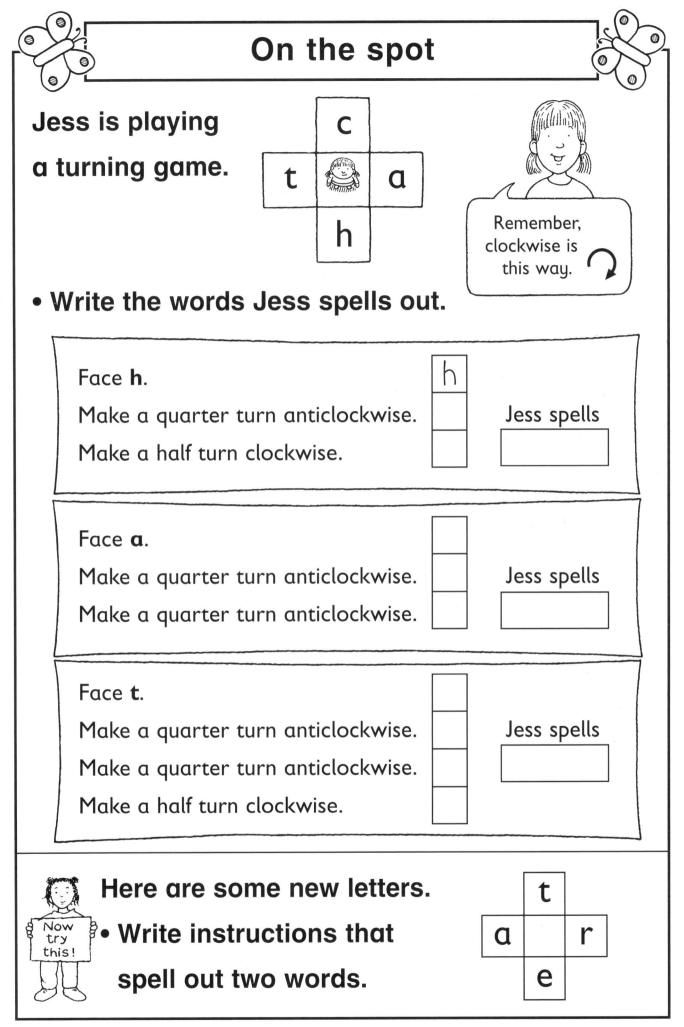

Jess is playing a turning game.

Remember, clockwise is this way.

• **Write the words Jess spells out.**

Face **h**.

Make a quarter turn anticlockwise.

Make a half turn clockwise.

Jess spells

Face **a**.

Make a quarter turn anticlockwise.

Make a quarter turn anticlockwise.

Jess spells

Face **t**.

Make a quarter turn anticlockwise.

Make a quarter turn anticlockwise.

Make a half turn clockwise.

Jess spells

Now try this!

Here are some new letters.

• **Write instructions that spell out two words.**

Teachers' note Encourage the children to notice that the same result comes from a clockwise or anticlockwise half turn. Further grids can be made, for example using the letters t, n, o, e to spell words such as one, ten, not, net, ton, note and tone.

**Developing Numeracy
Measures, Shape and Space
Year 2
© A & C Black**

Right angle gobbler

• **Make a right angle gobbler like this.**

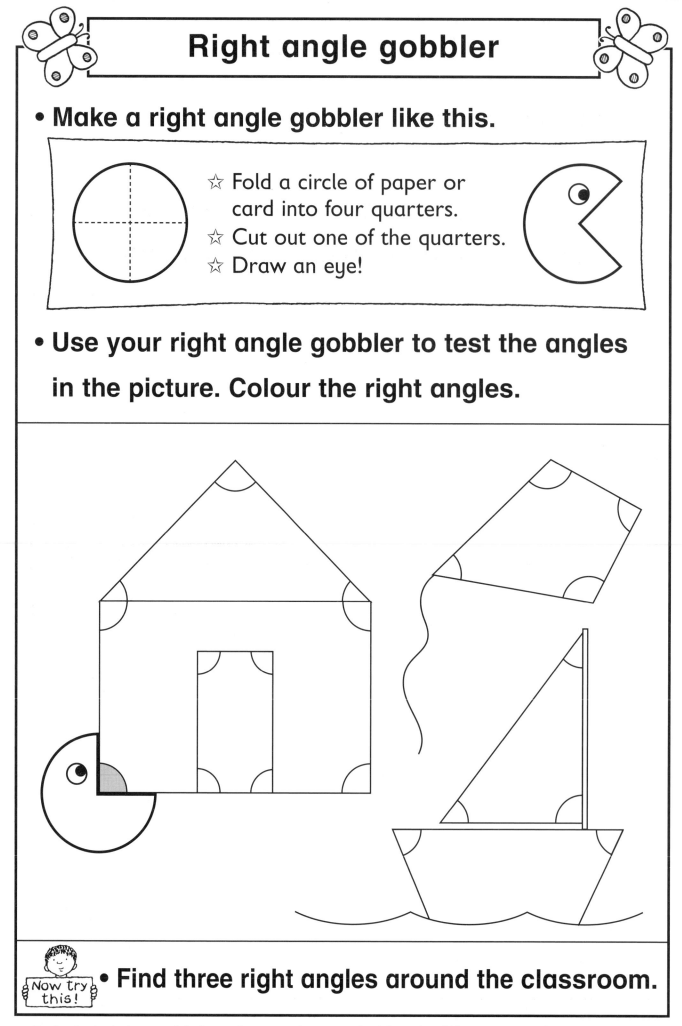

☆ Fold a circle of paper or card into four quarters.
☆ Cut out one of the quarters.
☆ Draw an eye!

• **Use your right angle gobbler to test the angles in the picture. Colour the right angles.**

• **Find three right angles around the classroom.**

Now try this!

Teachers' note At the start of the lesson, demonstrate how to use the right angle gobbler to test angles. Ensure that the children realise that they need to line up one edge of its 'mouth' with a line or edge and make sure that where the two lines or edges join is at the corner of its 'mouth'. If made with card or thick paper, these can be used for testing angles in the classroom.

Developing Numeracy
Measures, Shape and Space
Year 2
© A & C Black

Barmy borders

These shapes are turning clockwise through quarter turns .

- **Cut out the shapes at the bottom. Use them to help you draw the next two shapes in each pattern.**

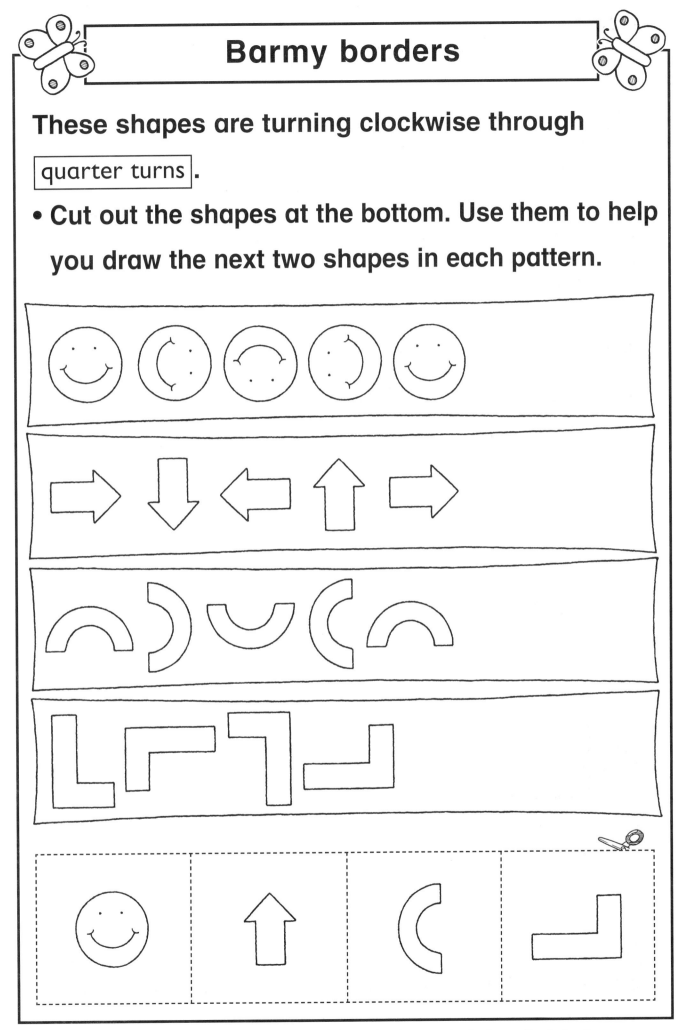

Teachers' note As a further extension, the children could use long strips of paper and a simple cardboard template or printing block to make their own large borders. Descriptions of the turns can be written to go alongside each border.

Developing Numeracy
Measures, Shape and Space
Year 2
© A & C Black

Tick the turn

• **Tick the correct turn.** ☑

> A quarter turn is the same as a right angle turn.

1.

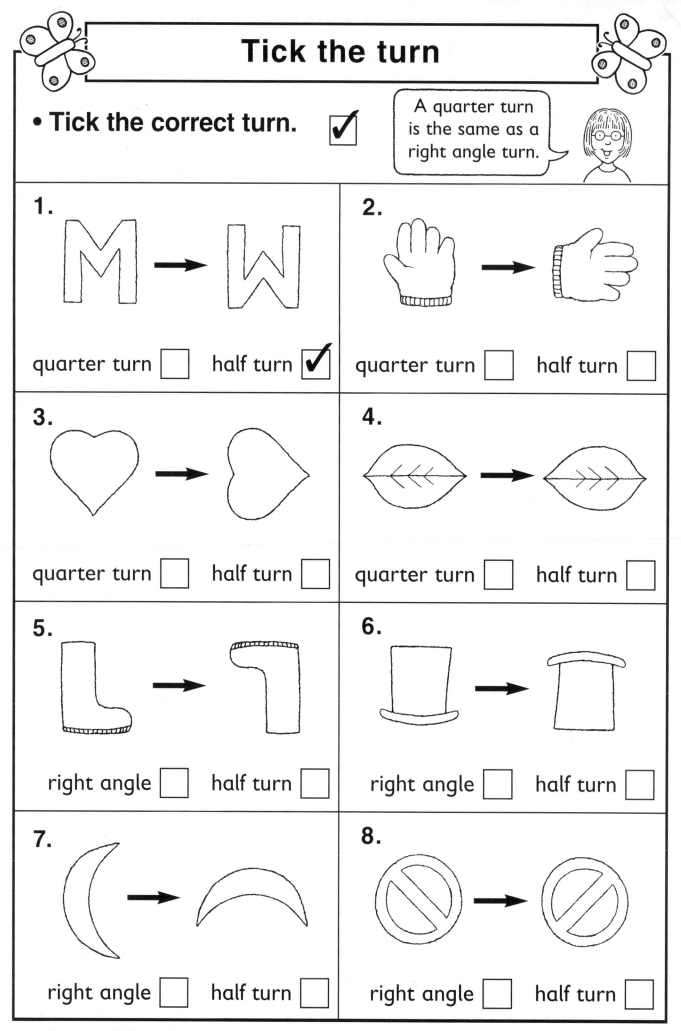

quarter turn ☐   half turn ✔

2.

quarter turn ☐   half turn ☐

3.

quarter turn ☐   half turn ☐

4.

quarter turn ☐   half turn ☐

5.

right angle ☐   half turn ☐

6.

right angle ☐   half turn ☐

7.

right angle ☐   half turn ☐

8.

right angle ☐   half turn ☐

Teachers' note This sheet includes the terms 'quarter turn' and 'right angle'. Discuss that turning through a right angle and through a quarter turn are the same. As an extension, the children could make a simple cardboard template and draw a repeating pattern where shapes are turned through a right angle or through a half turn.

Developing Numeracy
Measures, Shape and Space
Year 2
© A & C Black

Name the turn

Each object has made a turn.

• Write [half turn] or [right angle turn] .

> A right angle turn is the same as a quarter turn.

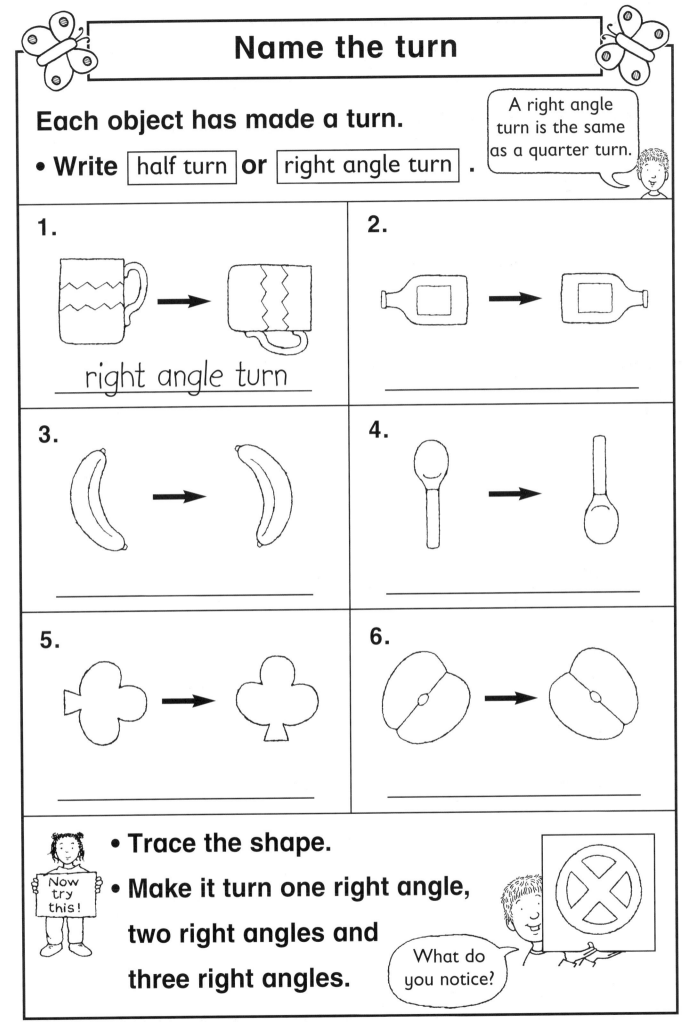

1.

right angle turn

2.

_____

3.

_____

4.

_____

5.

_____

6.

_____

Now try this!

• **Trace the shape.**

• **Make it turn one right angle, two right angles and three right angles.**

> What do you notice?

Teachers' note This sheet includes the term 'right angle'. Explain that this is the same as a quarter turn. As an extension, the children could make a simple cardboard template and explore shapes or patterns that remain the same after turns of one, two or three right angles, for example a square.

Developing Numeracy Measures, Shape and Space Year 2 © A & C Black

Beautiful borders

Always turn the shape in a **clockwise** direction.

- **Trace this shape.**

Use it to help you continue the patterns.

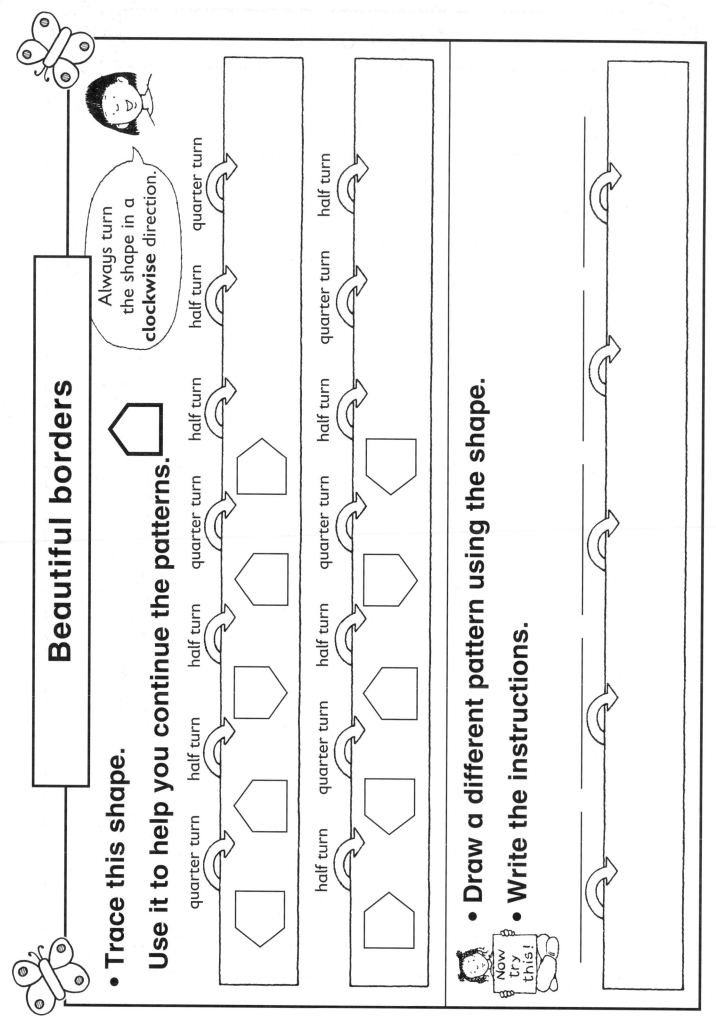

- **Draw a different pattern using the shape.**
- **Write the instructions.**

Now try this!

Teachers' note Some children may be able to complete this activity without tracing the shape. Others may need to trace the shape to experiment with different turns.

Developing Numeracy
Measures, Shape and Space
Year 2
© A & C Black

Acrobats!

• Work with a partner.

☆ Cut out the cards. Spread them face down.

☆ Take turns to pick a card. Put them all in a line.

☆ Each time, say how the acrobat has turned.

☆ Use the words quarter turn, half turn, or whole turn.

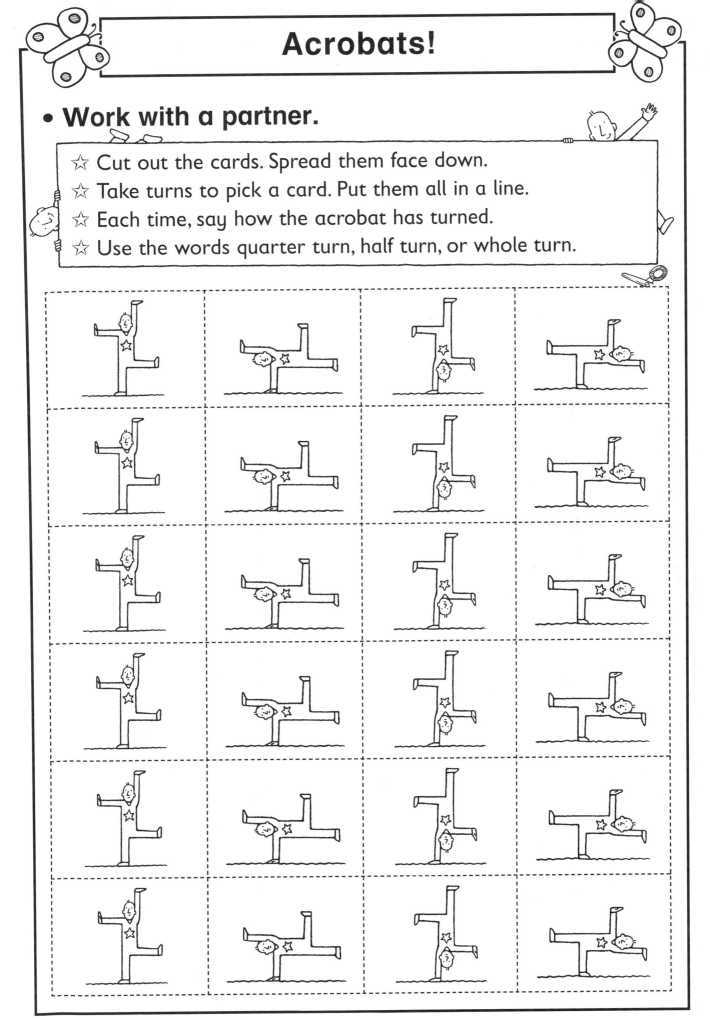

Teachers' note This sheet can be copied onto card and laminated to provide a more permanent resource. Alternatively, individual children can create and describe their own rotating patterns and stick the pieces of paper into their books. Use the cards to explain a 'whole turn' to the children and, if necessary, revise a 'half turn' and a 'quarter turn'.

Developing Numeracy
Measures, Shape and Space
Year 2
© A & C Black

Answers

p 7
Now try this!
caterpillar **b**

p 8
1. metres
2. centimetres
3. centimetres
4. metres
5. centimetres
6. metres
7. metres
8. metres
9. centimetres
10. centimetres

p 9
Metre stick, trundle wheel, ruler, metre stick or tape measure, trundle wheel, metre stick or tape measure, trundle wheel, ruler, metre stick or ruler.

p 10
Heights:
robot = 40 cm, sheep = 80 cm, teddy = 60 cm, plant = 70 cm
Lengths:
toddler = 90 cm, mouse = 10 cm, train = 50 cm, dinosaur = 30 cm

p.11
1. 5 cm   2. 7 cm   3. 4 cm   4. 8 cm   5. 9 cm

p 12
10 cm, 14 cm, 6 cm, 8 cm, 7 cm, 11 cm

p 16
1. kilograms
2. grams
3. kilograms
4. kilograms
5. grams
6. kilograms
7. grams
8. kilograms
9. grams
10. grams/kilograms
11. grams
12. kilograms
13. grams
14. kilograms
15. grams

p 21
1. millilitres
2. litres
3. litres
4. millilitres
5. millilitres
6. litres
7. litres
8. millilitres
9. millilitres
10. millilitres

p 23
Now try this!
b, i, a, d, g, f, h, c, e

p 25
1. October   2. May   3. June   4. December
5. July   6. December   7. January   8. September

p 33
1. sphere   2. cylinder   3. cube
4. cone   5. cuboid   6. pyramid

p 35
1. 12 edges 8 corners   2. 8 edges 5 corners
3. 12 edges 8 corners   4. 6 edges 4 corners
Now try this!
1. cube   2. pyramid
3. cuboid   4. pyramid (or tetrahedron)

p 36
1. true   2. true   3. true   4. false
5. true   6. false   7. true   8. false

p 38
1. 8 triangles, 5 squares, 9 pentagons, 10 hexagons, 5 octagons, 2 circles
2. rectangles

p 39
A and L are rectangles, B and F are triangles, C and G are pentagons, D and K are hexagons, E and I are squares, H and J are octagons
The shapes with 4 sides are: A, L, E, I
Now try this!
circle

p 40
1. 1 rectangle, 3 triangles, 3 pentagons

p 41

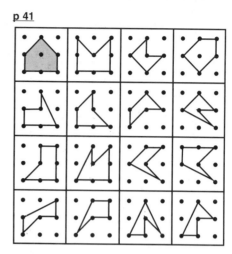

p 42
rectangle	octagon	hexagon	rectangle (square)
hexagon	octagon	octagon	decagon
hexagon	hexagon	octagon	decagon

p 43
triangle	pentagon	hexagon	octagon
hexagon	octagon	hexagon	octagon
pentagon	pentagon	octagon	hexagon

p 50
There are various possibilities. Check that the choice of word/phrase is appropriate.

p 51
Question 1, 2, 5 and 6 should be ticked.

p 52
C   L   T
U   S   H
Now try this!
E and G

p 53
1. hippo   2. giraffe   3. monkey   4. kangaroo
5. lion   6. elephant   7. panda   8. tiger

p 55

star 1 to star 2	star 2 to star 3	star 3 to star 4	star 4 to star 5
right 3	down 2	left 1	right 3
up 2	right 2	down 1	up 1
left 3	up 2	right 1	right 1
up 3	right 2	down 1	up 1
right 2	down 2	left 2	right 1
		down 1	up 3

Now try this!
down 3, left 1, down 1, left 1, down 1, left 3

p 56
1. bed   2. desk
3. TV   4. window
5. bed   6. TV
7. TV   8. TV

p 57
hat, act, that

p 60
1. half   2. quarter   3. quarter   4. half
5. half   6. half   7. quarter   8. quarter

p 61
1. right angle turn   2. half turn   3. half turn
4. half turn   5. right angle turn   6. right angle turn